Richard Adams

NATURE
DAY AND NIGHT

Illustrated by David Goddard

Science texts by Max Hooper
Illustrated by Stephen Lee

KESTREL BOOKS

KESTREL BOOKS
Published by Penguin Books Ltd
Harmondsworth, Middlesex, England

First published in Great Britain 1978

ISBN 0 7226 5359 X

Colouring Research by Jane Hunter and Shelagh McGee
Leaf-print on endpapers by Stephen Lee
Designed and produced by the Felix Gluck Press

Photosetting by Pierson LeVesley Ltd, Oxshott
Printed in Great Britain by W.S.Cowell Ltd, Ipswich

CONTENTS

To
CAROL ANN STORY,
δακρυόεν γελάσασα
Iliad vi. 484

Introduction

We human beings living on this planet find it natural to think and talk of Night and Day rather as we think about things which we can touch and see. "It was afternoon," we say, "and it was raining." And when we go to sleep at night, losing consciousness which we regain upon waking next day, our natural feeling is that there has been a break in the flow of time and that yesterday is separated from today by the intervening night and our own absence in sleep. It is strange and perhaps rather daunting to realize—really to feel and comprehend—that this is wrong. Night and Day are not "real" at all, in the sense that the rain is real: and time has been flowing on continuously, unbroken, since the beginning of the world. We live on a sphere of molten and cooled rock, crusted with earth, vegetation and water, which rotates on its axis and revolves round the sun. The part of it facing the sun at any particular time is warmed and lit, with a duration and intensity which varies because the earth is round, tilted in relation to the sun and revolving round it elliptically. The part not facing the sun is in shadow—part of that starlit dimness filling the entire universe, except those tiny parts where there happens to be a nearby source of relatively bright light. There is no such thing as Night and there is no such thing as Day. There is only this intangible covering of light and shadow, into and out of which any point on the earth's surface is continually turning. And there is no break at all in time, no division between one day and the next. When it is day in New Zealand it is night in England. If you had a fast enough aeroplane you could remain in perpetual daylight. That would, of course, be expensive, exhausting and pointless; but there are people who find it perfectly practicable, by travelling, to live in perpetual summer, which is what Ariel, in *The Tempest*, said he did.

> On the bat's back I do fly
> After summer merrily.

Even the solitary moon is an accident. If you lived on Mars you would look up and see two moons, and Jupiter has twelve.

So Night and Day are not "real", Science tells us. But in our hearts—in our bodies, our imaginations, in the very core of our lives and the lives of all creatures and plants on earth—there it is another matter. Night and Day shape and order our being, our work, our play, our tales and dreams, our feelings of boldness and of fear. Life has often been compared by poets, painters and musicians, to a dance. When they talk about "the dance of life," they mean that life is a matter of rhythms. We tend to feel active and excited in spring, eager and energetic when the sun shines in at the bedroom window in the morning. We may find that we feel oppressed and jumpy in thundery weather, and relaxed and sleepy after nightfall. The earth we live on is imposing its rhythms on us, and to feel and respond to them, each in his own way (for not everyone responds in the same way), is a vital part of our life on earth. The Irish poet, Padraic Colum, writing about *Grimms' Fairy Tales*, spoke of peasant life being governed by the rhythm of the sun and the rhythm of the fire. The first, said he, was a work rhythm: the second was the gentler, man-made rhythm of story-telling and listening.

Today these age-old, natural rhythms are everywhere being broken—by artificial light, by speed of travel, by altered patterns of work and play—in short, by Man's new-found power to change the world. Although man-made changes often seem so plainly for the better, we need to be careful how we interfere with the deep and very old rhythms that hold our world and ourselves in balance.

This book is about a paradox—Night and Day. They don't exist—we can alter them, or even turn one into the other if we want to. And they constitute the strongest power controlling the nature and rhythm of our lives.

Richard Adams

The Meadow by Day

Almost my earliest memory is of a meadow in June. The meadow isn't there any more—it's been built over—but the memory is. I must have been about five years old. A friend of mine had come to tea. It was midsummer and we had gone out to play in the long grass. I remember the big daisies—the ox-eyes—which were almost as tall as we were, and the fox-tail grass that you can pull out of its sheath and chew. Jean gathered buttercups, pulled off the petals and mixed them with the red-green flowerets of the sorrel which come off in nice, big handfuls when you pull your closed hand up the plant. "And what do you do with it?" I asked, looking at the red-and-yellow mixture—yellow wings in grainy, red powder. "You throw it at people," answered Jean. And we did. To make flower confetti in a mid-summer hayfield—not a bad way to spend an hour, at five years old. That was more than fifty years ago. I wonder whether Jean remembers? Another magical flower—or so it seemed then—that grew in that meadow was the orange hawk-weed (*Pilosella aurantiacum*). I loved its multiple flower-head, its hairiness and, of course, its wonderful orange-red colour—unique among English wild-flowers, I rather think.

There were grasshoppers in that meadow, too. The sound of their stridulations on a hot afternoon still brings it back to me. I used to creep very quietly up on them, to try actually to see them rub their files and scrapers together. You can't observe exactly how they do it, but as you hear the sound you can see that they are touching their wings and legs together. It still seems wonderful to me that a tiny, frail wing and leg can make so penetrating a noise. The green grasshoppers were pretty; and the brown ones glided so beautifully when they hopped, and had rosy wings which opened, so that you caught a glimpse before they folded them again upon alighting. Whenever I thought the grasshopper population of our meadow was falling too low, I used to spend an afternoon on the nearby common catching another hundred or so, which I released in the meadow that evening.

Nowadays, when I wander along the edges of a meadow at midsummer, it will usually be the birds that engage my attention. My great favourites are the warblers, and of them in particular, the blackcap and garden warbler. They sing best in the very early morning, about sunrise or even before, but on a fine June day they may be heard at any time. They like a high, thick hedge, scrubby bushes or an open patch of trees. The willow warbler, too, has a pleasant little song, (a "dying fall"), melancholy and quiet; not to be compared with the other two, but nevertheless a cool, green, leafy sound.

Then there are the finches; some common, some less common. Everybody knows the chaffinch, with his slate-blue head and wings flashing white in flight. Chaffinches can become very tame and will sometimes come and peck crumbs from almost under your feet. Even more delicate and fine are the goldfinches, with their scarlet, black-and-white heads and black-and-yellow wings; though I associate them with gorse and waste land rather than with true meadows, where they like to forage among tall weeds for thistle-seed and such-like food. The linnet, too, is often to be found along the meadow-side, particularly if the meadow is under the downs or near an upland.

But there are tougher gentry than these. The greenfinch, who sometimes looks almost like a canary in his conspicuous yellow-and-green, is a heavy, burly sort of bird for his size. He has three songs: the first is simply the repetition of a note with pauses between—"whit: whit: whit". The second is to speed this up—"whit-whit-whit-whit-whit". The third is his really characteristic call, a very rapid trill. "There's that greenfinch tearing his silk again," my father used to say. He didn't mind the greenfinch. His real wrath was reserved for the bullfinches; and indeed they are destructive birds, heavy-billed, with a kind of surly deliberation about them as they sit and pick bud after bud off a prunus in early March, or worse still, off the fruit trees. All the same, it is difficult not to admire them, with their splendid black skullcaps and vivid, pink breasts. A friend of mine, a fine ornithologist, used to say, in his north-country accent, "Y'know, life's better when ye've see a *bull-finch*, isn't it?" (He pronounced "Bull" to rhyme with "skull". It seems correct that way to me, now.)

Daisies and clover, sorrel and ragwort—the world would be greatly improved if these were part of everyone's summer memories. If you pull the flower head out of a red clover and suck the base, you get a taste of honey. Don't despise the ragwort either. If you crush the foliage, it smells beautiful.

"The insect world amid the suns and dew." That line is by John Clare and it shows he was a true poet, for he wrote "suns and dew" and not "sun and dew". Anyone could have written "sun and dew", but "suns and dew" takes you straight down into the long grass, don't you think?

Key to Illustration on pages 12-13

PLANTS AND TREES

1 Silver birch 2 Crack-willow 3 Alder 4 White water lily 5 Water crowfoot 6 Bindweed 7 Orange hawkweed 8 Great willow herb 9 Brooklime 10 Water figwort 11 Marsh marigold 12 Water violet 13 Ox-eye daisy 14 Bramble 15 Dog rose 16 Purple loosestrife 17 Reed 18 Red campion 19 Meadow buttercup 20 Lady's smock 21 Ivy 22 Honeysuckle 23 Bracken 24 Water forget-me-not 25 Soft rush 26 Meadowsweet 27 Water mint 28 Water cress 29 Reedmace 30 Wood sorrel

ANIMALS

31 Otter 32 Little owl 33 Blackbird 34 Moorhen 35 Garden warbler 36 Bullfinch 37 Goldfinch 38 Chaffinch 39 Linnet 40 Mute swan 41 Blackcap 42 Greenfinch 43 Whitethroat 44 Yellowhammer 45 Willow warbler 46 Song thrush 47 Common frog 48 Brimstone yellow 49 Small tortoiseshell 50 Gatekeeper 51 Mayfly 52 Cockchafer 53 Great diving beetle 54 Whirligig 55 Green tiger beetle 56 Pond skater 57 *Meta segmentata* 58 Garden cross spider 59 *Misumena vatia* 60 White-lipped banded snail 61 Great ramshorn snail

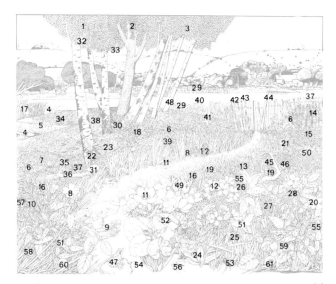

Moisture and Dew

Day and night seem different, because one is light and the other dark. The sun rises in the morning and sets in the evening—so we say. In reality, it is the spinning of the Earth on its own axis which gives us the sensations of day and night. The sun also provides warmth. It is warmer by day than at night and the warmth affects the amount of moisture in the air. When air warms up, moisture vapour is formed. But when air is saturated with moisture and then cools down, the vapour cools and is 'precipitated' as dew. You can test this for yourself by holding a cold plate near the spout of a boiling kettle. Steam comes out of the spout in the form of vapour. But when the steam hits the cold plate, water is formed, and this is exactly what happens in making dew.

Of course, if there is little water around, then the daytime air, however warm, will not create vapour. So during the day, the air may be drier than at night, when the small amount of moisture the daytime air has converted to vapour will be more than the cooler night air can support.

As a general rule, we can say that days are usually bright, warm and dry, and that nights are dark, cool and moist.

During the day, as the air warms up, the moisture in it is converted to vapour; at night, however, the air temperature drops, and the water vapour in it is 'precipitated' as dew.

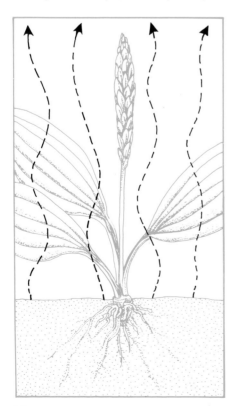 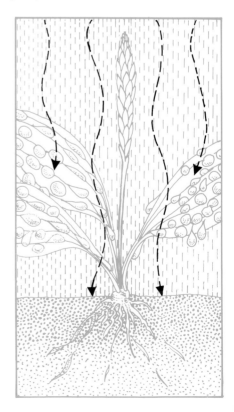

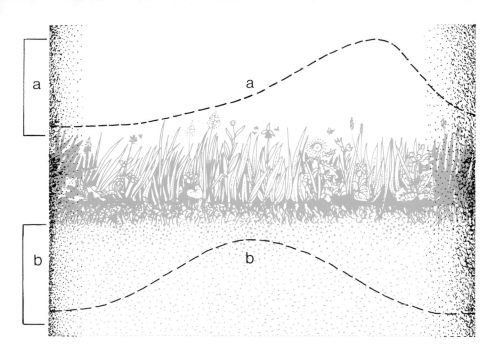

Soil is directly warmed by the sun. The air above it is warmed by the rising from ground level of air which in itself is warmed by the soil (a). As the sun descends after noon, so the soil cools (b), but the cooling soil is still heating the air above it, so the air reaches its highest temperature by late afternoon.

In every twenty-four hours, because of the constant spinning of the Earth, there are the bright, dry periods and the dark, moist periods— day and night. These affect all the many creatures and plants that we know so well and seldom think about.

Why do birds have a dawn chorus? Why do butterflies fly by day and most moths by night? Some of the answers to questions like these are easy. The tawny owl hunts by night because its food is active at night. The kestrel flies by day because its food is active by day. But why then is the prey of owls or hawks active by night or day? It is light and dark, or warmth and cold that brings them out. We depend on our eyesight rather than our sense of smell. But many animals depend on their noses. The fox, like so many of the dog family, has a keen sense of smell. He usually hunts at night. Are there differences in the senses of daytime hunters and night hunters?

Experiments have shown that many plants and animals have the ability to measure time—as if they had an 'internal clock'. Every day, at the same time, even in a laboratory where they are kept under the same light and in the same conditions of moisture and temperature for several days, they will perform the same activities. But the internal clock does not completely control the life of the organism. As long ago as 1836 it was noted that during an eclipse of the sun (at 2.15pm), bees which had been feeding in a meadow went back to their hive, and at the height of the eclipse an hour later, their hive was still quiet. Similarly, crows went to roost and midges swarmed, as if it were dusk. In this case, the darkening of the sun by the moon overruled the internal clock.

Nature's Clock Pattern

The sparrow-hawk preys on small birds, including the song thrush. It hunts by day, whereas the tawny owl is a night hunter, preying on small rodents, such as the wood mouse.

The song thrush is one of the earliest birds to be heard in the morning, can be observed during the whole day and will sing until late evening.

The wood mouse is active at night. It feeds on nuts and the occasional beetle, but by day it hides itself in its burrow—usually under a tree between the roots. The tawny owl (right) is one of its predators.

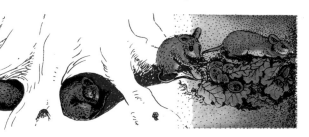

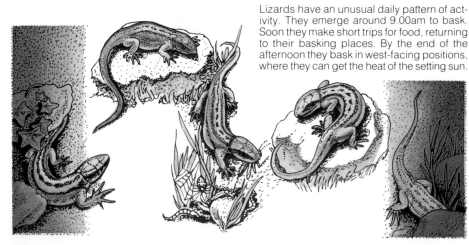

Lizards have an unusual daily pattern of activity. They emerge around 9.00am to bask. Soon they make short trips for food, returning to their basking places. By the end of the afternoon they bask in west-facing positions, where they can get the heat of the setting sun.

Clover (top) and wood sorrel have rather similar leaves, and they both close up at night, though in different ways as illustrated here. The reason for closing up may be due to the need to conserve heat.

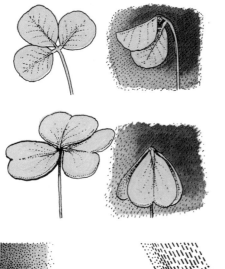

Dandelions, like their relations the daisies (day's eyes), close at night and in heavy rain, to protect their delicate stamens.

Scarlet pimpernels only open wide in full sunlight. At night or in cloudy conditions or when it rains, they remain closed. They are known as the Poor Man's Weatherglass.

The night-flowering catchfly is related to the red campion. Like the evening primrose (below) it opens at night, when it is visited by moths, such as the elephant hawk-moth. These night-opening flowers often have a distinctive scent, which attracts the insects.

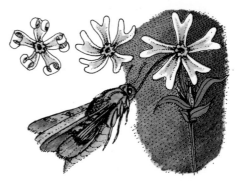

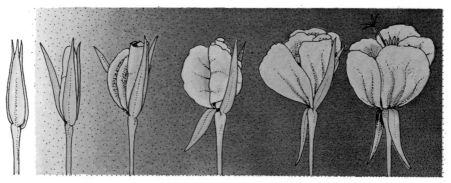

The Sky by Day

The sky is not naturally blue. Its apparent blue is due to the effect of the sun's light upon the envelope of air—the atmosphere—in which the earth is covered. To an astronaut travelling beyond the atmosphere, the sky appears black and in this blackness the stars seem to flare rather than twinkle. Our blue sky, above its clouds —white or grey, red, yellow, pink or black—is unique to the earth. It is not an optical illusion, any more than is a rainbow; but like a rainbow, it is a phenomenon brought about by the sun.

While we are still very small, before we can read or write or tell the time, we probably pay more attention to the sky than when we are older, for we have not yet learnt that looking at the sky is nothing but day-dreaming. Almost everyone can remember, at one time or another in early childhood, lying on his or her back, looking up and wondering what was beyond the sky and where it stopped—reflections that get you nowhere. And while we still have plenty of time to spare for things like looking for pictures in the fire and watching the rain or the snow fall— before we start being in a hurry to get on, or being afraid that other people may laugh at us for being silly—we often look at clouds to see shapes in them. Looking at clouds is idle—a waste of time, if you like—and is only done, for the most part, by people who, either because they are very young or very old, or because they have nothing more important to do with their time, either for the moment or for ever, are not troubled about wasting it. Shakespeare knew this. On the one hand he knew that clouds and the sky are fascinating, while on the other he knew that most people feel, naturally and without reflection, that anyone who stops to look at the clouds is idling, at a loose end. This is why Hamlet, when he is deliberately trying to annoy the old statesman Polonius and also to deceive him into thinking that he is mad, compels him to hang about talking about the shapes of the clouds. What more rubbish could one talk than that? But Hamlet is himself a time-waster. This is his tragedy. And again, the great emperor Antony, when he learns (falsely) that Cleopatra is dead; that his empire is in ruins and his time has come to die, begins to speak to his squire of the changing, forever-vanishing shapes of clouds. It is a marvellous stroke of dramatic poetry; for, in the first place, as Antony himself says, not only has he himself, with all his glory, become like a mere cloud-shape about to melt away, but secondly, we feel unconsciously the pathos of the defeated general, who until this had all the world's cares on his shoulders, but now can find no more to do than look at—or at any rate talk about—the shapes of clouds.

And yet shapes are important and have meaning in themselves, as every artist and sculptor knows. The shape and mass of a fruit, a stone or a cloud have significance and meaning needing no explanation beyond themselves, and we are the poorer if we do not feel this importance. Looking at the sky and the clouds—being conscious that the sky is above us as the earth is beneath us—is not a waste of time if it gives us a sense of the sky's simplicity, grace, vastness and purity. There is as much skill required to paint the sky above a landscape as to paint the landscape itself. It is no good just painting in a few clouds of no particular shape or form and saying that that is what clouds are like anyway—all a matter of chance. If you doubt this, go and look at some of the great landscapes—Turners, Constables, Monets or Sisleys—in the London galleries or elsewhere, and notice how much care the painters took over the clouds and the sky, and what such care does for a picture. The sky is complementary to the earth, always. Earth and sky are in balance.

To be aware of the nature and quality of the sky above the country round us, to realise how it affects the character of that country, and to compare it, and the particular kind of light it sheds, with other lights in other places—all this is an important and exciting aspect of our life on the earth. The thick, still, creamy blue of a hot August sky, which seems to lie only just above a wood of beeches, is different from the remote, thin blue of a fine October afternoon, through which the sun seems to float as though down a current of clear water. The wet, cloudy skies of Ireland shed on that green-hedged country and its little, stony fields a very different light from that which falls upon Norfolk, where mile upon mile of ploughland stretches from one horizon to another. I well remember, some years ago, a young South African girl, who had been in England only a few weeks, saying that to her, the most striking difference between Johannesburg and England was the extraordinary subtlety and variation of the English light. "In Johannesburg," she said, "there is only one kind of light. Here the different lights seem infinite."

The sky and its clouds are a country of fantasy and daydream, of time spent to no purpose but leisure and the play of the eye with the imagination. Like sleep and its dreams, it is always there and always making its effect upon us, whether we regard it or not. I think that what Proust said of the relationship of sleep to human life could as truthfully be said of the relationship of the sky to the earth. "One cannot properly describe human life unless one shows it soaked in the sleep in which it plunges, which, night after night, sweeps round it as a promontory is encircled by the sea."

Key to Illustration on pages 20-21

CLOUDS

1 Cumulus 2 Stratocumulus 3 Stratus 4 Nimbostratus 5 Cirrus

PLANTS AND TREES

6 Hard Fern 7 Heather

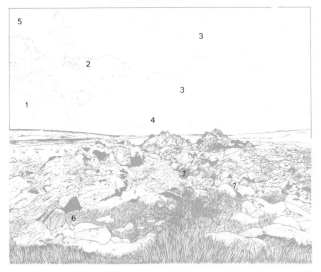

Hot Air and Cold Air

One of the best ways of looking at plant and animal life is to think of it as a struggle for survival—both for the individual organism and for the species as a whole. Only those which survive to reproduce can be called successful.

The surroundings (or environment) in which any organism lives are of first importance. There will be a complex mixture of danger and opportunity; but at least parts of the mixture are constant: day follows night; winter is succeeded by spring. Many organisms, with their internal clock, can predict these changes, and this makes their lives easier, and safer.

Apart from day and night and the seasons, many other changes are more or less predictable—such as wind, weather, temperature and dampness. On bare ground, temperature is at its lowest at sunrise. It rises rapidly to its maximum by noon, and falls quickly towards sunset, and then more slowly towards dawn.

When areas of high pressure in mountains coincide with low pressure in the lowlands cold air above the mountains may become channelled down the valleys as the warmer air in the lowlands rises. Strong dry winds may then result, blowing down the valleys at considerable speeds. Well-known winds of this kind are the mistral, blowing from the Alps to the Mediterranean, the bora of the Adriatic coasts and the chinook of the Rocky Mountains.

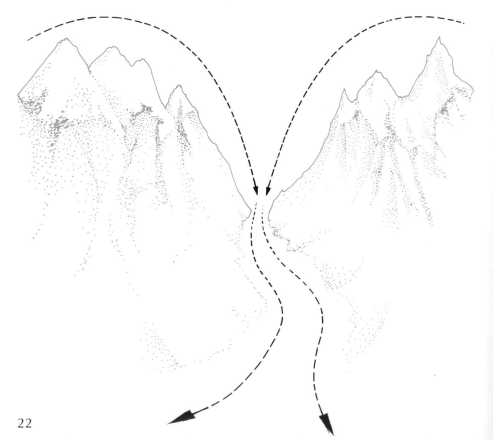

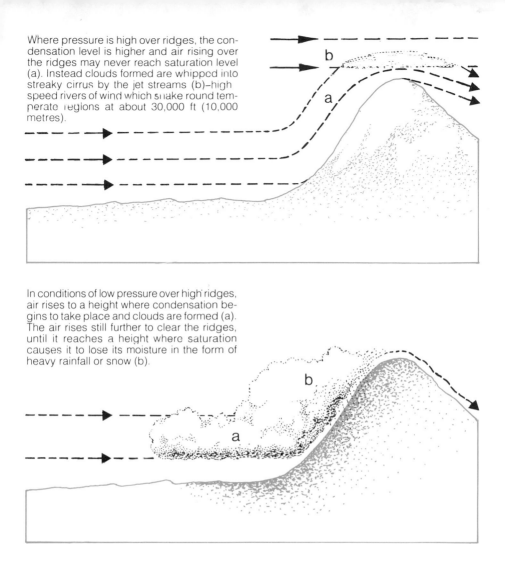

Where pressure is high over ridges, the condensation level is higher and air rising over the ridges may never reach saturation level (a). Instead clouds formed are whipped into streaky cirrus by the jet streams (b)–high speed rivers of wind which snake round temperate regions at about 30,000 ft (10,000 metres).

In conditions of low pressure over high ridges, air rises to a height where condensation begins to take place and clouds are formed (a). The air rises still further to clear the ridges, until it reaches a height where saturation causes it to lose its moisture in the form of heavy rainfall or snow (b).

The air also follows a similar pattern, but it warms up more slowly than the soil, and is at its warmest in early afternoon, for it is the soil which heats it rather than the sun. However, there are complications. A dark peat soil warms up quicker than a white chalk soil, and peat is often damper than chalk. This means that since water loses its heat more slowly than soil, peat may be constantly warmer than chalk. And if we compare any soil with water in a stream or in a lake, we will find that because the water warms up more slowly and loses its heat more slowly, the difference in temperature between day and night is less than in soil.

As air itself is warmed, it gets thinner and rises. From noon to sunset breezes tend to run uphill; at night, as they cool, they become heavier and run downhill.

Warm air can absorb more moisture than cold air. Since warm air rises, it eventually is cooled, and the moisture it carries condenses and forms clouds. This is why on a warm summer's day there are often heaped-up cumulus clouds and rain on the mountains.

So even non-living things like clouds and rain follow a daily pattern.

Cloud Formations and Weather Charts

a) Cirrus—high clouds, with more or less dense heads from which streaks or tails fall.

b) Cirrocumulus—very high heaped clouds in dappled sheets, associated with 'mackerel skies'.

c) Cirrostratus—very high shapeless clouds, sometimes covering the whole sky.

d) Altostratus—medium high clouds, associated with 'watery' sun. These clouds cover many miles of sky with thickening layers of haze.

e) Altocumulus—medium high heaped clouds, not so high but similar in dispersal to Cirrocumulus.

f) Stratus—low or very low shapeless clouds on hills and coasts.

g) Cumulo nimbus—the typical towering thunder clouds, which can rise up to Cirrus level.

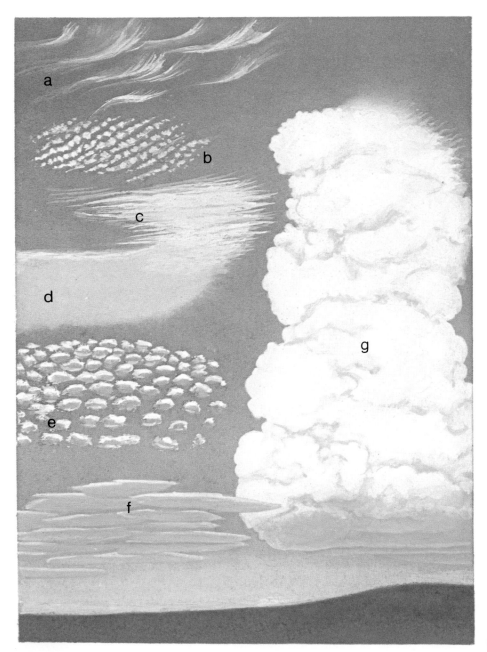

Weather and Weather Charts

	Isobars (millibars)
	Cold front
	Warm front
	Occlusion
	Cold surface winds
	Warm surface winds

In the accompanying weather chart, two fronts—a warm (to the right) and a cold (to the left)—are shown. A warm front occurs where warm air has replaced cold air by sliding over it; a cold front occurs where faster-moving cold air has cut under warm air, forcing it upwards. Depressions are formed when a small bend is created in a more or less stationary front, with warm air on the inside of the bend. As the bend becomes more pronounced, pressure falls and wind increases. The cold front moves faster than the warm front and overtakes it and, where it overtakes, an 'occlusion' is created (at top of this chart). Finally the cold air has moved all the warm air from the centre of the depression, and the front becomes more or less stationary again.

HIGH

LOW

The Meadow at Nightfall

I am walking very slowly, with many halts and pauses, along the banks and hedges of a meadow at sunset. It is August—a perfect evening, with a red sunset promising another scorching day tomorrow—and this is Berkshire, in the south of England. Whatever am I doing? I am looking for a nest of the common wasp (*Vespula vulgaris*), and if I succeed in finding it, I shall poison it, since it is far too close to the house and the kitchen in particular has become almost impossible. Wasps' nests grow bigger as summer goes on, and these particular wasps, wherever they may be hanging out, have become a nuisance and a danger. Who wants to be stung? The queen, the only wasp who survives the winter, founds her nest in spring, nearly always underground, making paper from wood pulp chewed from trees, fences and so on with her powerful jaws. By August, a nest may contain 20,000 marauding wasps. If, however, you go out at sunset and wander quietly about, you are likely to become aware, after a time, that there is an intermittent but steady stream of wasps, all going one way. They are workers obeying their instinct to return to the nest at sunset and, unlike a bird or animal, they cannot deceive you or lead you elsewhere. Like all insects, wasps have no free will and can only obey their instincts. But if you do come upon a nest (all you will see will be a small hole in the bank with wasps going in and out) take care! They may very well fly at you and sting you, even if you are doing no more than to watch them going in and out. Nevertheless, to dig out a "dead" nest entire is fascinating and well worthwhile. It is spherical, and may have as many as eight tiers of cells, each covered round with a thin sheet of paper. I stand in wonder before the incomprehensible miracle of wasps' organisation; and I know, too, that they eat and destroy many other harmful insects. But all the same, as Mr Churchill said, up with them I cannot put.

While I am wandering along I meet another toiler of the dusk—a hedgehog. He is scuttling his way along the ditch, in and out of the brambles and cow parsley, looking for slugs, worms and beetles, all of whom are more active at dusk, because they prefer the coolness and dew. I've a good mind to carry him back to the garden, for a resident hedgehog is a godsend to any gardener. He is the real leprechaun for whom you put out a saucer of milk at night to induce him to stay and work for you. However, I leave him alone. No doubt he prefers his own locality. Just for fun, I touch him gently, to see him roll up. His spines are very sharp indeed —you have to be extremely careful how you pick him up in your bare hands. Also, like all hedgehogs, he is full of fleas under his prickles. They have their use, for since he can't wash himself, like a cat or rabbit, they keep him clean in the soft fur under all those prickles. The hedgehog is very much a nocturnal hunter. If you find one walking about by day, the poor fellow's generally dying. If you do put a hedgehog in your garden and make a pet of him, don't molest him with too much attention. Little creatures—mice, moles, toads, hedgehogs—are timid, and if they are continually pestered may actually die as the only way of escape.

There are more night prowlers now, for it is after sunset. I catch glimpses of bats fluttering silently against the paling sky. These are pipistrelles, the commonest kind, but there are long-eared bats in the neighbourhood too, and I keep a lookout for them. Bats hunt moths and other insects at dusk. They are marvellously nimble flyers and very intelligent. Once, when one flew into my bedroom on an August night, I shut the window so that he couldn't escape. He flew all round the perimeter of the room three times—once at floor level (including under the bed), once at intermediate level and once at ceiling level—without colliding with anything. Having satisfied himself that there was no way out, he went to sleep on top of the cupboard. I let him go later, of course. (They have fleas too.) While you are still young, you can hear their tiny, high cries, (*twink, twink*) but as one gets older the eardrum becomes less sensitive. I shall never hear bats again.

I have just found and dealt with the wasps' nest when suddenly, quite close by, I hear a tawny owl call from the oak tree. The call is in two parts; the first, "H-h-h-hoo"; then a pause of about four seconds; then "Huh. H-h-h-h-hoo". Some say it is ventriloquial, to deceive mice and voles as to where he actually is. It probably does have the effect of frightening them into doing something silly, for otherwise he would hardly advertise his presence so clearly, especially as he is totally silent otherwise. A moment later I see him flying over the hedge, right across the huge, rising harvest moon.

My way home takes me past an old, thick bush of yellow-leaved holly. In here, among the piles of dead leaves and snug and dry under the live ones, lives a colony of field crickets (*Gryllus campestris*). As the heat of the day cools, I can hear them stridulating away in their prickly castle. Good luck to them—crickets have always been regarded as lucky.

Over the hedge from the garden steal the smells of phlox and night-scented stock. On the tennis-netting just inside the field gate sits a poplar hawkmoth, laying eggs by the hundred. It seems an odd place for her to be laying. Never mind —let her alone. The enormous moon is clear of the horizon-haze now, turned from ruddy orange to silver. Sirius, the dog-star, is bright too at this time of year. That's why they're called the dog-days. And Jupiter is shining so bright as almost to cast a shadow. A far-off clock strikes nine, the sound floating faintly up from the town below. Could a night be more beautiful?

Key to Illustration on pages 28-29

PLANTS AND TREES

1 Bay willow 2 Crack-willow 3 Alder 4 Silver birch 5 Pedunculate oak 6 White water lily 7 Marsh marigold 8 Reed 9 Water crowfoot 10 Bindweed 11 Dog rose 12 Ivy 13 Water cress

ANIMALS

14 Water shrew 15 Hedgehog 16 Bats 17 Swift 18 Pochard 19 Redshank 20 Tawny owl 21 Little owl 22 Common frog 23 Green lacewing 24 Drinker moth 25 Convolvulus hawkmoth 26 Poplar hawkmoth

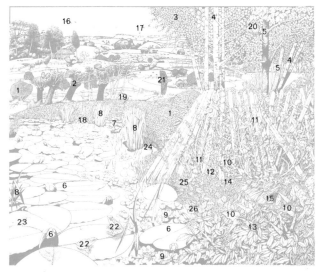

Nocturnal Patterns

As evening falls so does the light and the temperature of both the air and the soil; but the moisture level in the air rises. The change in temperature and moisture level of the air is most marked just above the level of the grasses; but deep down amongst the tufts of grass the air is still relatively warm and moist. Insects become active as the light fades. Woodlice and beetles begin to forage among the tufts of grass, crickets begin to call, and as the humidity rises the scents of the evening get wafted further in the moist air. Moths begin to appear, midges swarm, and bats begin to catch their nightly food supply. Among plants, meadow grasses open their flowers. First in the late afternoon is meadow fescue, then couch grass and sheep's fescue; later still meadow grass itself. Mice and voles begin to feed amongst the grass, and the tawny owl arrives to hunt them. Snails and slugs appear and the hedgehog to hunt them.

But is it light, heat or moisture which starts up these activities? Or is it the internal clock, or indeed is the cause something else not directly related to any of these?

Sometimes the answer is clear: slugs, snails and many insects would lose too much moisture during the day: their activities do not depend on light and can be carried on at night when the atmosphere is damp and they are less likely to dry out. But then why should mice, voles and shrews come out in the evening, only to be harried by the owl? Why should the hedgehog come out at night?

These are all animals which as a group have developed powers of sight and therefore, one might think, could perform all their activities efficiently by day. Why then have these creatures chosen to give up their apparent natural advantages?

For the hedgehog and the shrews the increased activity of slugs, snails and insects may be the answer—here is a rich source of food which other predators have not found. They are taking advantage of the situation. But what of the vegetarian mouse? The grass is there day and night; the mouse has eyes to see and can control his body temperature, and does not as easily lose moisture in the sun as a woodlouse might.

Perhaps it is because the owl is the lesser of two evils—that by day the kestrel is more dangerous than the owl by night.

Grasses (from left to right)

Meadows are active places by night and by day. While the grasses carry their flowers high up to enable the wind to carry their pollen, a myriad creatures are moving round their roots. (a) snails and slugs (b) woodlice (c) velvet ground beetles (d) crickets (e) centipedes (f) millipedes (g) earthworms (h) earwigs.

Meadow brome *Bromus commutatus* flowers in the morning and the evening
Couch grass *Agropyron repens* flowers in the late afternoon
Sheep's fescue *Festuca ovina* flowers in the afternoon and evening
Meadow grass *Poa pratensis* flowers at night
All grasses are wind pollinated

Luminescence in Nature

Some fungi, such as honey fungus *Armillaria melea* and a bracket fungus *Panellus stipticus*, actually produce a faint glow in the rotting wood off which they live. They are usually found growing on rotting bark and the blue-green glow of the honey fungus can only be seen when it is fairly dark. But this luminescence is far more marked in certain insects such as glow-worms.

Glow-worms: the winged adult male (left) and the wing-less female.

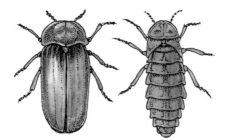

In the western tropics, fireflies are common. These beetles give off an even greater glow than the glow-worms. They are known to illuminate whole trees in the tropics.

The female glow-worm gives off a much brighter light than the male and its purpose seems to be to attract the male.

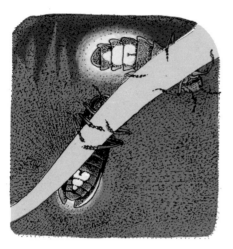

Fireflies fly in an erratic wavering pattern, which enlivens the evening sky with zigzags of light.

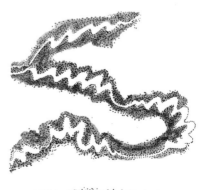

Glow-worms will feed off snails. Above, a female is injecting a snail with a fluid which dissolves the internal structure of the snail, enabling the glow-worm to suck out the contents.

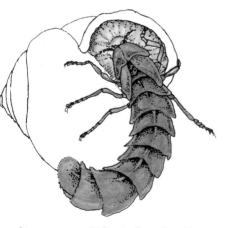

The Sky at Night

I am always surprised that most people pay so little attention to the stars. As often as not, someone whom you know to be fond of birds, wild flowers and the seashore will admit to being unable to recognise any constellation except the Great Bear and perhaps Orion's Belt, and doesn't seem to think he wants to do anything about it. Yet even a simple ability to identify the principal constellations (of either hemisphere), without knowing any more astronomy than that, can give great pleasure. As with any other sport or hobby, you get out what you put in. Once you have got a star map and learned the names of a few stars, the mere exercise of your knowledge is enjoyable: to pick out the faint North Star, either by projecting from the "Pointers" of the Great Bear, or else, if the Bear should be clouded over, by bisecting the angle between two stars in Cassiopeia; to look round a clear night sky and recognise the brighter stars—Algol in Perseus, Arcturus in Boötes, Vega in Lyra and so on—or to try, for the thousandth time, to count seven stars in the Pleiades (I suppose there must be seven, if they say so, but I can never see more than six with the naked eye)—this sort of thing gives a lot of enjoyment and satisfaction. Once you have really stood shivering and taken in Orion glittering on a winter's night—sword, dagger and belt, Rigel and Betelgeuse—you can never ignore it again. It becomes, as it were, part of your personal furniture, like snowdrops, or Long John Silver. Once, when I was pointing out the stars in Orion to an Arab friend, he told me (he couldn't help smiling) that in Arabic "Rigel" (he pronounced it "Rïjjel") means "toe". The civilised orientals of the Middle Ages (whom we know from *The Arabian Nights*) named the stars long before we in Europe knew much about astronomy, and the Crusaders and early scholars adopted the Arabian names, which we still use.

Don't get mixed up with astrology, which is pernicious rubbish, an insult to the dignity and beauty of the stars (rather like performing animals) and incidentally contrary to Christian teaching and belief. Having said which, I must admit that the Zodiac constellations provide a lot of sport. They always remind me of that limerick about the eggs at Devizes—"Some were so small, They were no use at all, But others won several prizes". The early astrologers needed names for those groups of stars which, for their purposes, "tied up with" the sun throughout the twelve months. This being so, they had to take whatever happened to be, during each particular month, in that part of the sky which was "in opposition". For this reason, only about five of the Zodiac constellations are really conspicuous and beautiful, the others being rather dim. In fact, you can hardly see Cancer at all, and Aries and Pisces aren't much better. But Taurus, Gemini and Scorpio are magnificent, while Leo, with his chief star Regulus—as seen from England, anyway—is my own favourite constellation. Most of the Zodiac constellations don't look much like their names, but Scorpio really does suggest, strongly, the shape of a scorpion. In the latitudes of the British Isles, Scorpio rises only a little way above the south-western horizon in high summer, and I had long been accustomed to think of its chief star, Antares, as a dull, glowing red. This, however, is merely the effect of summer heat and horizon haze, through which Antares's light has to travel to England. It was a bit of a shock, in the South Seas, to recognise Scorpio up in the zenith, with Antares glittering silver and almost as bright as Jupiter.

To look at the moon through binoculars, to perceive that it is curved, a sphere in space, and pick out its rather frightening, desolate hills and valleys: to cut out the monthly "Night Sky" column from *The Times* (or some other newspaper) and to look for some of the planetary and other appearances of which it gives notice; to know when meteors are due—such things are as exciting, and as much a part of Nature, as rocks, birds and flowers. One can, of course, if one wishes, go on to learn about supernovas, red dwarfs, black holes and all the rest of it—the point being that just at present this is a very important branch of human study, in which exciting advances are being made. Some really big breakthrough in human knowledge will almost certainly take place before the end of the century. But all I am saying here is that to know and to be aware of the stars adds another area of enjoyment to life. It does more than that. During the war, when I had to leave my home and those I loved, my way of life and all that had been familiar to me and to go abroad, as a soldier, to live a strange life among strangers, it often seemed that nothing—not the birds, not the flowers, not even the seasons of the year—remained the same. But the stars did, and they were a great source of comfort.

Key to Illustration on pages 36-37

STARS AND PLANETS

1 Corona Borealis 2 Hercules 3 Lyra 4 Cygnus 5 Draco 6 Ursa Minor/Polaris 7 Ursa Major 8 Cepheus 9 Cassiopeia 10 Triangulum 11 Perseus 12 Pleiades 13 Boötes 14 Mars

CLOUDS

15 Altostratus halo 16 Altocumulus

PLANTS

17 Sea couchgrass 18 Crested hairgrass 19 Sea spleenwort

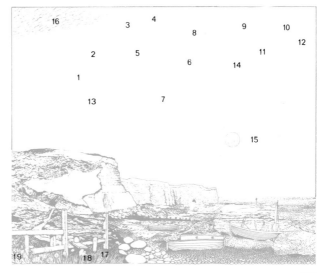

Animal Navigation

Amphipods, which are related to crabs and shrimps, are found throughout the seas at all depths, and also on beaches. One is called *Taliturus saltator*. If it is taken a short way up the beach, it will head back towards the sea. However, it is not just heading for the sound of the waves or for the sensation of moisture, for if it is taken to another beach directly across the bay it will head *inland*—in other words, it is following a compass bearing, not just the lure of the surf.

This creature appears to steer by the position of the sun. But since the sun does not stay in the same place all day, *Taliturus* must have an internal clock to work out in which direction to head.

And what happens when the sun is obscured by cloud? Insects can overcome this because their eyes are sensitive to some light which we cannot detect without special instruments. This is 'polarized' light, which is filtered through the atmosphere. Its angle of polarization changes according to the position of the sun. So even on cloudy days insects can find their way, using polarized light from the sun.

In spring and summer huge numbers of insect-eating birds fly north on migration from Africa and the Mediterranean to take advantage of abundant food supplies. Some well-known examples are the swift (a), the swallow (b), the reed warbler (c), and the cuckoo (d).

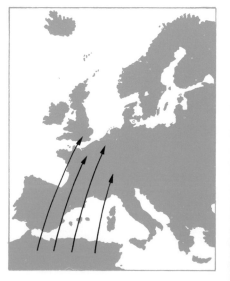

But what happens at night? Birds are the best known navigators. Even quite small birds, such as swallows, can fly accurately over enormous distances during their migrations, and come to rest at exactly the right place several thousand miles away. But swallows fly by day. Radar has shown that there are massive numbers of other small birds which also fly by night. They somehow have an internal sense of the right direction. How they do it is still a matter of discussion. It is probably done by a combination of navigating by the sun, the stars, and the magnetic fields of the Earth. But the sun and the stars move in relation to the Earth—except for the Pole Star in the Northern Hemisphere. So birds must have an internal clock to take account of these apparent movements. It has been suggested, and it is possible, that night-flying birds can actually recognize the Pole Star. Mallard (wild duck) will keep on a straight course by day or night. If they are given artificial daylight, their course will alter as though they were navigating by the sun using their internal clocks; but by night they keep their course steady. It would seem that mallard, at least, navigate at night by the Pole Star.

Other birds, living in harsh northern climates, move south during the winter. Examples are the fieldfares (a), the brent geese (b), and the grey lag geese (c). Woodpigeons, though resident in Britain, undertake winter migrations on the European continent from Sweden to the Pyrenees.

Stars and Constellations

The familiar stars at night are as reassuring to man as they are invaluable to migrating birds. Obviously, even on the clearest nights, only part of the sky can be seen. The Pole Star can be seen only in the Northern Hemisphere; the Southern Cross only in the Southern. Some constellations, such as Orion, lie low on the horizon in the Northern Hemisphere and can be also seen in the north of the Southern Hemisphere.

Scientists have classified stars at varying magnitudes which we have marked accordingly on this chart.

1st magnitude

2nd magnitude

3rd magnitude

4th magnitude

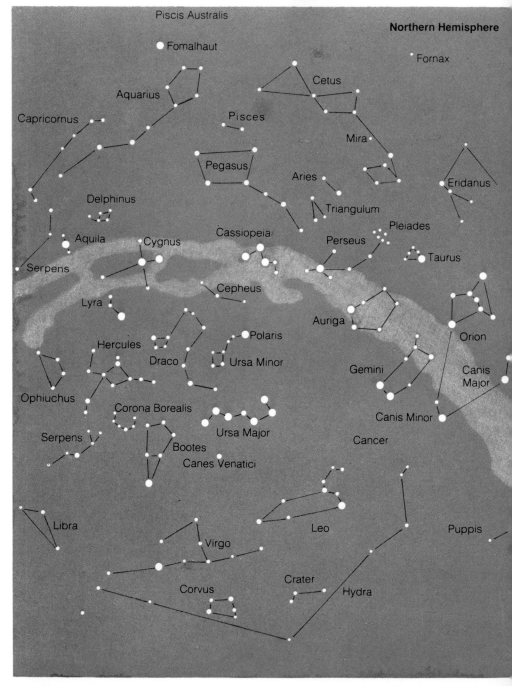

The Milky Way indicated by the lighter area is made up of millions of stars.

Groups of stars are called constellations. Many were given their fanciful names in Classical antiquity. Some of the best known in the Northern Hemisphere are Cassiopeia, Cepheus, Cygnus, Pegasus, Ursa Major and Ursa Minor; and in the Southern Hemisphere, Canis Major, Centaurus, Crux Australis (Southern Cross) and Orion.

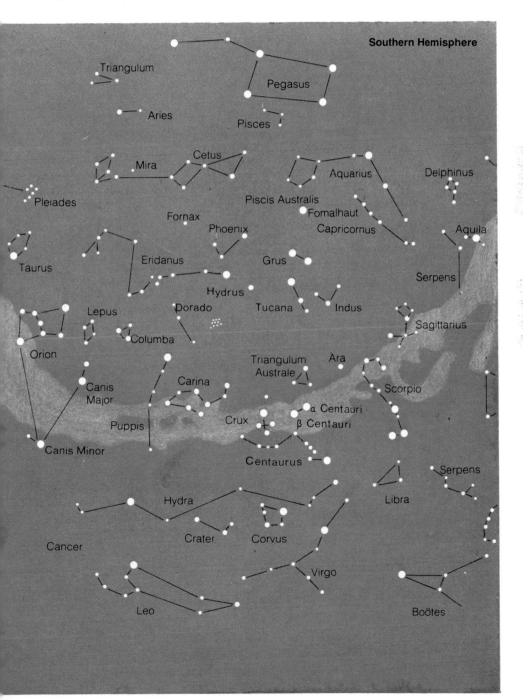

Southern Hemisphere

Triangulum

Pegasus

Aries

Pisces

Cetus

Mira

Aquarius

Delphinus

Pleiades

Piscis Australis

Fornax

Fomalhaut

Phoenix

Capricornus

Aquila

Eridanus

Grus

Serpens

Taurus

Hydrus

Lepus

Dorado

Tucana

Indus

Sagittarius

Orion

Columba

Triangulum Australe

Ara

Canis Major

Carina

Scorpio

α Centauri

Puppis

Crux

β Centauri

Canis Minor

Centaurus

Serpens

Hydra

Libra

Cancer

Crater

Corvus

Virgo

Leo

Boötes

The Moorland by Day

Of all kinds of country in the British Isles, moors —so it has always seemed to me —
are the most individual. Just as the deserts of the world vary in character (the Sinai
is not at all like the Kalahari, for instance), so Bodmin Moor differs from the York-
shire moors, and they again from the Lake district and the Scotch highlands. Then
there are the smaller moors—tracts of peaty heather—dotted about southern
England; such as the ones round Newbury, in Berkshire, which I happen to know
well —Greenham Common, Newtown Common and Snelsmoor. While each has
its own nature and characteristics and its own special features, most have many
birds and flowers in common. (Sorry!)

I am standing on a high, lonely moor in the south of the Isle of Man. It is noon on
a hot day in August. The bare hills, seen from a distance, are coloured a brilliant,
glowing purple by their covering of heather. This is the so-called bell heather
(*Erica cinerea*), the prettiest and most brightly-coloured. But even on this one
moor (from which I can see the Irish Channel, blue as the sky above it), several
different kinds are to be found. The ling (*Calluna vulgaris*) is like a tiny hearth-
brush, with smaller, paler flowers than the bell heather, and minute, evergreen
leaves. The cross-leaved heath (*Erica tetralix*),which tends to prefer boggy places
in the peat, is very striking, with globular flowers like little lanterns bunched in
tufts at the tops of the stems. White heather (which is not a separate species, but a
freak, like a white blackbird) you can, if you are lucky, find here and there; usually
small patches or individual plants growing among the ordinary, purple heather.
(The "white heather" that gypsies and others try to sell you is nearly always ling:
real white heather is less easily found.)

Bilberry plants (*Vaccinium myrtillus*) grow thickly on this Manx moor, but
August is usually too late for the delicious berries, which grow on the tiny, green,
dry bushes (for that is what they really are) in June or July, after the red, waxy,
globular little flowers are gone.

There are several strange and attractive moorland flowers. I particularly like the
sundews (*Droseraceae*)—the fly-catchers—whose long, syrup-haired leaves
radiate from the root to attract and catch tiny flies exactly as a fly-paper does.
When the fly is caught on the sticky leaf, it slowly closes on the poor fellow and the
plant absorbs food from him. The small, white flowers rise on long stalks from the
centre of the plant. Then there are red rattle and lousewort, which are prettier than
their names. The pink and crimson, hook-shaped flowers are usually not very hard
to find on moorlands between May and September. If this were May and we were
on a western Irish moor, we might hope to find the Irish heath (*Erica hibernica*);
and if it were July and we were in Scotland, we might look for the trailing azalea,
with its leathery leaves and deep pink flowers. Cranberries, too, you can find, in
the right place. It is all a matter of knowing your own moor. Apart from flowers,
this one has the added advantage of being near the shore and the sea.

And the birds? On the moors more than anywhere it is a great help to have
binoculars for the birds, for they can see you from a long way off. What have I seen
here today? Linnets, twites and yellowhammers, as one would expect, and rooks,
hooded crows and jackdaws. It is worth learning to tell these last three from one
another—not merely by their colouring but also by their behaviour; for often one
sees a bird only against the light, and four-fifths of the art of bird recognition is to
be alert to where you saw the bird, to what it was doing, how it was behaving or
flying and how many there were together. I was lucky enough to see a great,
"bearded" raven and I also saw a magpie, which perched, cocked its long tail and
flew away, flashing black-and-white like a very old film. What pleased me most,
however, was a whinchat. He may have been on migration south—"in passage",
as they say. He is a brown bird about as big as a chaffinch, with a white eyestripe, a
narrow white "collar" and a rather dowdy, plum-coloured breast. He perches

very upright, nearly always on the top of whatever he sits on—weeds or rocks—and is also to be seen hovering, with fluttering wings, just above grasses, posts, wire and the like. His cousin, the stonechat, is similar in behaviour, but has a very striking black head. I put up a blackcock, too—a black grouse—with a lyre-shaped, double tail and red wattles over his eyes. He went off like a rocket across the moor.

It is very hot and the heather and gorse smell delightful.

Little cowboy, what have you heard   *(that means, boy watching the cows)*
Up on the lonely rath's green mound?   *(a rath is a prehistoric earthwork)*
Only the plaintive yellow bird   *(the yellowhammer)*
Sighing in sultry fields around,
'Chary, chary chary chee-ee?'
Only the grasshopper and the bee? . . .
How would you like to roll in your carriage,
Look for a duchess's daughter in marriage?
Seize the shoemaker—then you may!

The poet, William Allingham, goes on to tell how he himself once came upon the fairy shoemaker, who escaped capture by flinging the contents of his snuffbox in Mr Allingham's face. If only one could catch him and hold him tight—so the legend ran —he had to give up his gold. I never found him myself: but I love the poem, which for me has all the hot solitude of the moorland on an August day, when the only sounds are those of the insects, the yellowhammer and linnet; and perhaps an occasional, distant sheep or two, if it happens to be that sort of moor.

Key to Illustration on pages 44-45

PLANTS AND TREES

1 Bog asphodel 2 Bog violet 3 Cloudberry 4 Trailing azalea 5 Red rattle 6 Marsh andromeda
7 Lousewort 8 Heath bedstraw 9 Ling 10 Bell heather 11 Cross-leaved heath 12 Purple moorgrass
13 Crowberry 14 Bilberry 15 Gorse

ANIMALS

16 Mountain hare
17 Rabbit
18 Weasel 19 Curlew
20 Ptarmigan
21 Red grouse
22 Common buzzard
23 Kestrel 24 Swift
25 Ring ousel 26 Raven
27 Twite 28 Grayling
29 Green hairstreak

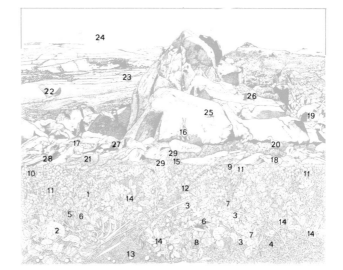

Leaf Movements and the Sun

At first sight, moors and heaths are not unalike. They have the same sorts of plants: bracken, heather, cottongrass, sedges and grasses—such as mat grass and hair grass. But there is one big difference: moorlands are wet, and heaths are dry, though they have their boggy patches. In Britain, heathland is found mainly in the lower, drier south and east of the country, on acid sands and gravels; moors are found in the north and west on higher ground where there is more rainfall.

The plants in both types of country are mostly adapted to avoid loss of water. They are constructed to cope with a harsh and dry environment.

Plants have many ways in which they can avoid drying out in the midday sun. They can reduce their leaves and their height to the minimum, so that less of their surface area is exposed to the sun—and this means that they lose less water by evaporation. Think of the leaves of heather, bilberry, gorse or broom. They are small or very small, with tough, leathery surfaces. Another way in which plants can prevent themselves drying out is by extending their underground system—their roots—so that they can give their roots the greatest chance to absorb as much moisture from the soil as possible. Many heath grasses do this: for instance, sheep's fescue.

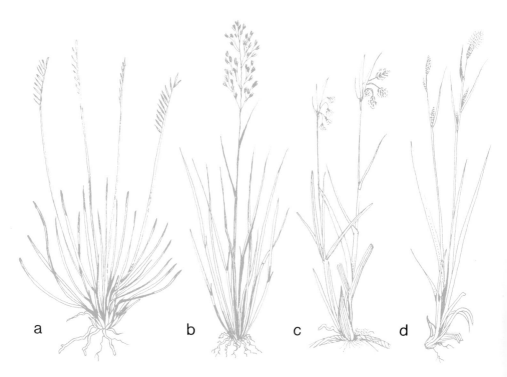

Common moorland grasses: (a) Mat grass *Nardus stricta* can be extremely abundant, since sheep will not eat its tough leaves. Hair grass *Deschampsia* spp. (b) usually grows in the damper spots. Cotton grass *Eriophorum* spp. (c) is very obvious in bogs where its silvery bristles can create a sheet of white on the moor. Sedges (d), of which cotton grass is one, differ in several ways from grasses, especially in having solid not hollow stems.

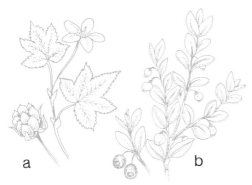

Heath and moorland plants are typically low-growing, often with small, tough, leathery leaves, to cut down loss of water through evaporation. The Plants shown here may be locally dominant.

Cloudberry *Rubus chamaemorus* (a) is like a low-growing blackberry with a few orange or red drupes. Bilberry *Vaccinium myrtillus* (b) is found both on open moors and in woods. Its fruit is delicious when cooked. Ling *Calluna vulgaris* (c) likes well drained soils. Broom *Sarothamnus scoparius* (d) and gorse *Ulex* spp. (e) both brighten the moors in summer with their golden flowers.

a   b

c   d   e

As has been noted, the leaf surfaces are often leathery, and sometimes waxy. This slows down evaporation, and some plants can even roll their leaves up to slow down the escape of moisture. Several grasses do this. They have small pores (called 'stomata') only on their upper surfaces. Plants breathe through the stomata, so that when a grass blade rolls up it protects the stomata and moisture is conserved.

Other plants can move their leaves away from the heat of the sun. Some have structures at the base of the leaf which easily lose water, and when this happens the whole leaf is unable to support itself and droops. When moisture in the air is high, these structures become full again of water, and the leaf rises again to its normal position.

You can see leaf movements quite easily in runner beans in your garden. The plant holds up its leaves during the day, but at night they droop. The runner bean seems to have a 'clock' inside it, because it keeps on raising and lowering its leaves even when it is kept in continual light or continual dark.

Spiders and their Webs

Spiders spin their webs and catch their prey in varying cunning ways.

The garden spider *Araneus diadematus* spins its beautiful symmetrical web to trap flying insects.

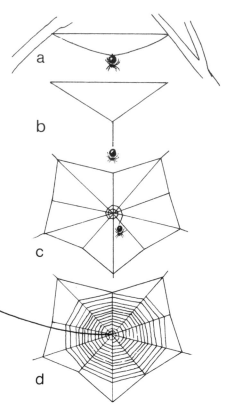

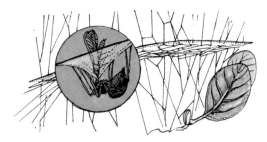

The little red money spider has a sheet web, and here is seen with an entangled greenfly (*Lynyphia triangularis*).

Above, the ways the garden spider spins his web.

(a) Usually a link between two branches is made by a spider floating in the wind.
(b) The spider drops from the centre of the link making a third connection with another branch or leaf.
(c) Other connections radiate away and a primary winding is made from the centre.
(d) Secondary final links are spun to trap insects.

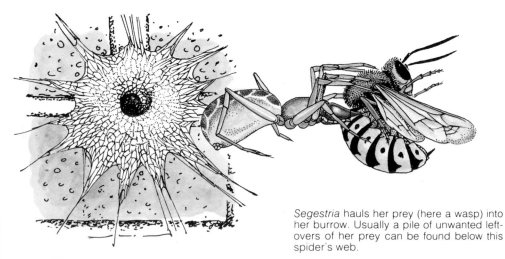

Segestria hauls her prey (here a wasp) into her burrow. Usually a pile of unwanted leftovers of her prey can be found below this spider's web.

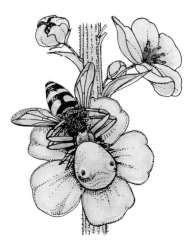

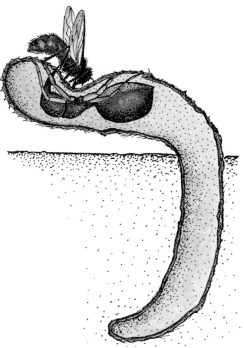

The crab spider *Thomasis*, seen here on dark mullein *Verbascum nigrum*, is camouflaged by its colour. Here it is pouncing on a visiting hover fly.

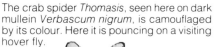

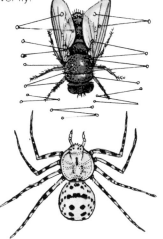

The purse web spider *Atypus* waits in a tube below ground, and pierces and immobilises insects passing over the surface of the tube.

Scytodes, the spitting spider, 'spits' a thread round the prey and wraps it up for later meals.

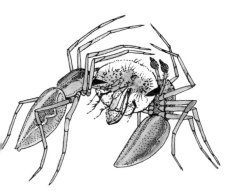

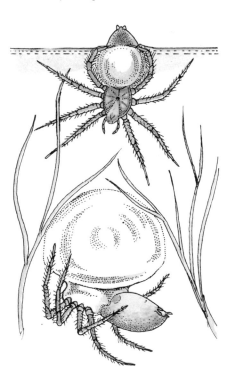

A hunting spider, *Pisaura*, presents its mate with a wrapped up fly.

The water spider, *Argyroneta aquatica*, constructs a diving bell between fine webs spun below the surface of the water in pond weeds. It feeds on aquatic larvae and water lice.

The Moorland at Nightfall

Moors at nightfall have always been thought of as sinister places. As everyone knows, Macbeth and Banquo met the witches on a "blasted heath". ("Blasted" here means "withered" or "barren".) Thomas Hardy, in his novels, several times introduces the bitter waste of Egdon Heath, often at dusk; and "the lonely moor" was a common feature of the "Gothick" literature popular about a hundred and eighty years ago, (which gave us "Frankenstein").

Though none is supernatural, I am certainly happening upon several very lethal creatures pursuing their business this evening, while walking on this lonely, heathery moorland a thousand feet above the sea—which lies still and dark-blue about three miles away. The first is a nest of ants (*Lasius flavus*, I think) whose domed, grassy mound of a nest, about as big as a good big basket, I should have passed without a glance if my attention had not been attracted by the inch-thick column scurrying purposefully into and out of the heather. Following it up, I come, about three or four yards from the nest, upon the body of a young rabbit, over which they are crawling in their thousands. It is bleeding from the mouth and one ear. The ants couldn't have done this, so either they finished it off or it died before its smell attracted them—the latter, I hope.

There are something like thirty-six species of British ants, all living in colonies and mainly carnivorous hunters. Carrion attracts them, but they can also kill for themselves—anything helpless. In fact, these ants are only doing their scavengers' job in devouring this poor young rabbit to the bones. As I kneel close above them to watch their extraordinary and continual exchange of information by rubbing heads, they become alarmed and angry, exuding their formic acid. The bitter, ammoniac smell is strong enough to make my eyes water. There's something disturbing about their world, in which every ant automatically fulfils its function and the lives of individuals count for nothing.

The dusk is deepening, and as I wander down a green path through the heather a white owl appears, flying low about twenty yards to my left. Even though it is growing dark, he probably wouldn't have come so close if I hadn't been concealed, as he approached, by the corner of a belt of pine trees. His wings flap slowly and make never a sound. He passes unhurriedly, deliberately turning on me for a moment the big, dark eyes in his white-downed face; a brief, stony gaze.

Here's something he missed, though—possibly because of me—a male stag-beetle (*Lucanus cervus*) a good two inches long, crawling slowly round the base of a rotten tree-stump. I pick up his shining, hard, brown-black body between finger and thumb, and he waves his legs in slow motion. He looks terrifying, with those huge, antler-like mandibles, but is actually quite harmless, his jaws being too large for the muscles that control them, so that he can't inflict a painful bite. These beetles seldom move around much, just clambering about near their source of food—usually plant roots or rotten wood. I put him on my hand, admiring his complicated, delicate legs, and wait to see whether he'll split open his carapace, shake out his wings and fly—a splendid sight if only he'll do it. However, he remains torpid and unco-operative, so I return him to his tree-stump.

It's getting darker than ever as I come up to the narrow, lonely road which I mean to follow home down the hill. I'm standing still to look at the last ochreous streaks of sunset out at sea when suddenly, about sixty feet away, a polecat-ferret emerges from the heather and halts doubtfully on the verge. He's a big animal, a good foot long, dark-masked and dark-eyed, but the rest of him creamy yellow, the fur long and coarse like a ferret, and with a very noticeable tail. He trots quickly across the open road and disappears.

These polecat-ferrets are an interesting feature of the Isle of Man. The Island has a few stoats but no badgers, foxes, otters or weasels. (Weasels are common on most other British moorlands.) The polecat-ferret is, therefore, almost the only

Manx predator capable of hunting and killing the beautiful Scottish hares (*Lepus timidus*) which abound on the hills. (Unlike the big brown hare, *Lepus europaeus occidentalis*, these turn pure white in winter.) Having been brought to the Island many years ago to hunt rabbits, the polecat-ferrets escaped and multiplied in the wild. They are deadly killers and a great nuisance to farmers for, like foxes, they will often kill in a kind of feral frenzy, destroying far more than they need or can carry off. An adult polecat-ferret can tackle geese and turkeys almost as easily as hens (or kittens) and is very fierce when cornered. Altogether, a thoroughly nasty bit of work. Nevertheless, the sight of one setting out across the moor to his night's hunting is exciting—rather like meeting a highwayman or sighting a pirate ship. I can't help feeling on his side. (I don't keep chickens, of course.) He reminds me of Sredni Vashtar in Saki's story about the boy who worshipped a polecat-ferret which finally killed his horrible guardian.

It's quite dark now, the sky merely a paler darkness above the black trees on the edge of the moor. Against it, small bats are flittering silently, hunting for moths and other insects. I have a soft spot for bats and admire the skill with which they fly and hunt. Even among thick trees, they never touch or blunder into obstacles, partly on account of the sensitive whiskers round their muzzles, but mainly because their minute, high-pitched squeaking acts as a kind of radar, echoing in their great, delicate ears from nearby objects and thus enabling them to perceive and avoid them. They snap up their prey in flight and usually pouch and eat it without alighting. They hunt all night in summer; and even during hibernation, winter sunshine warm enough to awaken insects will often bring them out too, to hunt by day. There are twelve different kinds of bat in Great Britain, but these little fellows are just pipistrelles, the commonest.

I conclude that moors don't need ghosts or witches. There are quite enough fear-dealers and killers up here, without resorting to the supernatural. In the wild every creature is part of the food cycle, either a killer or a victim—often both—and there's no such thing as death from old age. This grim aspect of nature is not—to put it mildly—easy to accept, but to ignore it is to evade the truth and therefore to understand Nature less clearly. Killing, eating and survival—that's what it's all about.

PLANTS AND TREES   Key to Illustration on pages 52-53

1 Bell heather 2 Marsh gentian 3 Deer grass 4 Common cotton grass 5 Star sedge 6 Common sedge 7 Bog willow herb 8 Heath rush 9 Spotted heath orchid 10 Moss campion 11 Yellow mountain saxifrage 12 Starry saxifrage 13 Mossy saxifrage 14 Hairy stonecrop 15 Hard fern 16 Sphagnum 17 Lichens

ANIMALS

18 Polecat-ferret
19 Mountain hare
20 Brown hare
21 Pipistrelle bat
22 Short-eared owl
23 Emperor moth

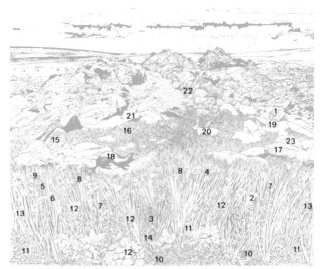

Sense and Sound

We find it more and more difficult to see things as night falls. Yet with animals, they often become more active at dusk and during the night. How have they overcome the dimness of the light? Some have managed by developing other senses—senses of hearing, smell, and special modifications to their eyes which enable them to see in the half-light of dusk.

The eye has an area called the 'retina'. This is the part of the eye which is sensitive to light, and it contains two kinds of sensitive cells, called 'rods' and 'cones'. Rods and cones have different functions. Cones see colour and detail, while rods pick up movement and detect small changes in the intensity of light. So by finding out the proportion of rods and cones in an animal's eyes, you can get some idea of how it lives. A predator, such as a kestrel, a merlin or a buzzard, needs to pick up tiny tell-tale movements on the ground as it hovers or soars over the moor. These birds have a high proportion of rods.

The light-sensitivity of the retina is increased in some animals by a reflective layer, which lies behind the retina, called the 'tapetum'. This strengthens the dim image which the eye sees in the dusk. You may have seen the reflection from the tapetum if you have ever seen a wild animal, such as a fox, caught in the headlights of a car.

Some moths, such as the emperor, have a highly developed sense of smell, enabling them to find their mates over quite long distances.

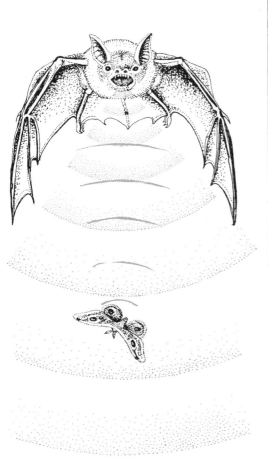

The sensitivity of the eyes of some animals is increased by the presence of a 'tapetum', a reflective layer behind the retina. It is reflection from the tapetum which makes a fox's eyes glow bright red in the headlights of a car at night.

Bats find their prey by echo-location. High-pitched squeaks are continuously emitted and bounce off flying insects, enabling the bat to track them down.

But obviously, if an animal which hunts by dusk or night and has very sensitive eyes is brought into daylight, it may be dazzled by the brightness. This is also overcome. Most animals have a third eyelid. We have one too, but it is tiny and is not used. However, in owls, for example, this translucent lid is drawn across the eye in daylight and acts very much in the same way as sunglasses do for humans—it cuts down the dazzle.

When it is totally dark, eyes are almost useless, so the truly nocturnal animals rely on other senses. The most remarkable, and the best known, is the sense of hearing developed by bats. When bats fly they continually squeak. Short-wavelength sounds uttered by the bats bounce off objects and are reflected back to the bats' sensitive areas.

Moths have evolved mechanisms to defeat the bats. Some can give out high-pitched squeaks, like the bats', which confuse the bats' 'radar' system; others are so furry that the reflected squeaks are indistinct; others dive for escape by twisting and turning, after detecting the bats' high-frequency emissions.

Bats can also distinguish between obstacles (such as trees or walls) and prey (such as insects).

Animal Vision

Ultra-violet light. A flower which to man might appear a uniform yellow, may have ultra-violet pollen guides visible to the bee.

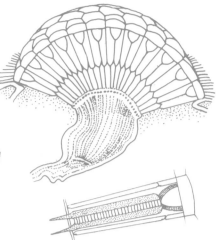

A cross-section of an insect's eye shows it is made up of many different lenses. Each lens unit picks up a part of the scene which registers in the insect's brain and produces the compound image.

Hawks, such as the kestrel or sparrowhawk, have eyes set well to the side of the head. This gives them very wide vision, though they cannot see directly in front of them. They have more cones than rods to ensure good daytime vision and excellent perception of details.

Owls' eyes are set in front of their faces, but cannot be moved. As a result owls have to move their whole head to change their direction of sight. The large pupils enable them to see better in dim light, and their eyes are also sensitive to infra-red which enables them to spot the warmth coming from small animals in the dark. They have both rods and cones, but a stronger density of rods in the central retina gives them great sensitivity to light, enabling them to hunt at night.

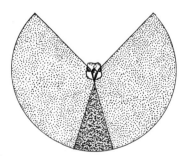

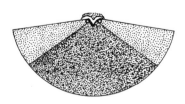

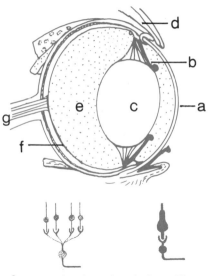

Cross-section through typical eye. The rods and cones are light-sensitive receptors embedded in the retina. (a) cornea; (b) iris; (c) lens; (d) muscle; (e) humour; (f) retina; (g) optic nerve.

Eyes of different animals are sensitive to different colours, from ultra-violet to red. Below we show the extent of their colour vision.

(a) man
(b) bumble bee
(c) falcon—bird of prey
(d) frog—amphibian

Frogs appear to be attracted to blue. This may help them to escape from aquatic predators, since the blue of the sky attracts them to the surface. Their eyes are also adapted to seeing above water while the body is submerged.

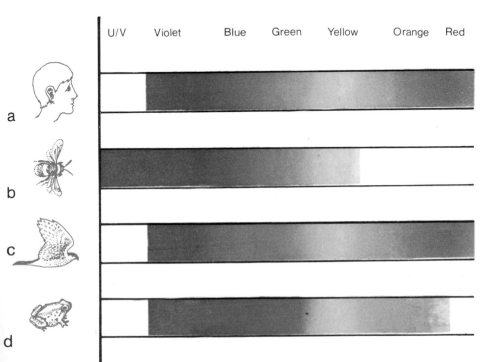

	U/V	Violet	Blue	Green	Yellow	Orange	Red

a

b

c

d

The Woodland by Day

This wood where I am walking, in the west of England, extends over either side of a steep, fairly wild valley. Its outer edges, which stretch right up to the moor above, are a new plantation—350 acres of larch, pine, fir, spruce and hardwoods, not yet twenty years old, tall and thick, densely set and almost impenetrable. The Norway spruce is what we all know as the Christmas tree, but these cannot have been planted for that purpose, for already they are far too tall.

I have seen both the tree creeper and the nuthatch on the outskirts of this dense and rather gloomy forest. Neither are big birds. Both live by picking insects out of the bark of trees, but their ways of going about it are different. The stumpy-tailed, pointed-beaked nuthatch (slate-blue above and chestnut underneath) climbs round and round and up and down the tree-trunks in a nervous, jerky way, holding on with his long, flat-splayed claws. The tree creeper, which has a brown back, a white breast and a rather long, downward-curving beak, also has splayed claws, but he almost always climbs only upwards (or along the branches), flying down when he thinks he has gone high enough. He flutters about, seldom flying far— unlike the nuthatch, who is perfectly capable of darting away and out of sight. Both birds are rather shy, but if you live near a wood you can often attract them to your bird-table, especially in hard winter weather.

There is little point in pushing into the heart of the plantation's twilit dimness. Insects don't penetrate there much, so the birds don't either. One interesting thing I notice as I skirt the edge, on this afternoon of late September. On the ground near the larches is a number of large, yellow-brown toadstools, their broad caps rather wavy, moist and shiny on top. Underneath, they look like the inside of a crunchie bar. This is *Boletus elegans*, which is found only near or under larch trees. It is perfectly safe to eat and tastes pleasant.

I prefer, to the coniferous plantation, the deciduous wood at its heart, which lies on either side of the stream running down the centre of the valley. This wood, as one can tell from the trees, is much older. There are huge oaks, ash trees and syca-mores, the latter with their leaves all covered with the black patches of the so-called tar-spot fungus (*Rhytisma acerinum*). A green path winds up through the wood, along the right bank of the stream. The trees are so big that there are plenty of open places between them, and here the forester (I wonder why?) has evidently tried his hand at planting a few less common trees. There are some Scotch laburn-ums —or could they be Voss's laburnum, I wonder?—with their graceful, pendent, triple leaves. And here is a low, spreading tree covered with clusters of bright-red berrries among the oblong, dark-green leaves. Whatever can it be? It turns out to be a Himalayan tree-cotoneaster. The berries have a bit of competition, for the rowans are blazing orange while along the edges of the thickets hang the dark-red berries of the thorn trees and the lighter, rugger-ball-shaped wild rose hips. There is a thrush pottering about, but evidently he isn't interested in the berries yet. That he leaves them alone now is a sign of a hard winter coming, or so they say. The birds leave the berries on the trees during autumns when they know they are going to need them later. Birds know and do so many inexplicable things that it could very well be true.

High up in this old wood the channel of the little, tumbling stream—no more than four feet across here—has been lined on either side with dry-stone walling, strong and nearly five feet tall, so that the water runs in a kind of miniature gorge, overgrown with black-berried elders. Nearby stand the ruins of a stone-walled cottage. This was once a mill: not a busy mill, used all the year round, with a resident miller and horses coming and going. It was one of those small, occasion-ally occupied mills used long ago by remote communities as opportunity offered. Often this stream is almost dry between its stone-walled banks, but after a day of heavy rain off the Irish Sea it comes pouring and rushing down, full of the heavy,

58

brown spate off the moor above. At such times the villagers used to seize their opportunity, hurry up the valley and get their corn ground while there was sufficient head of water to turn the wheel. Wheel and stones are long gone now and only the little, ruined house remains above the beautiful, neat, strong walling along the banks. It has been a hot, dry summer and the stream is running low, deep down between them.

On the wall of the ruined mill an unusual insect is walking—a kind of beetle. He is rather smaller than my little fingernail and brown, with six orange legs, a pointed "nose" and a brown-backed body shaped like a tiny shield. Towards the base of his back, in the middle, is a single, clear-red spot. I look him up in the book and he turns out to be a forest bug, *Pentatoma rufipes*, found among trees. These are one of the numerous family of shieldbugs, not found much in England, it seems, except during or after hot summers. The book says he does no harm and it's certainly more his wood than mine, so I leave him where he is and stroll home, picking ripe blackberries off the brambles as I go.

Just as I am leaving the wood I see a rather striking plant growing out of the stony bank. It has completely round, slightly scallop-edged leaves, with the stalk in the middle underneath, very cool and fleshy and most of them rather bigger than a 50p coin. Each leaf has a round dimple in the upper-side centre. Above the leaves, the flowers form a little spike or tower of greenish-white bloom, but these are nearly over now and withering. Some call this pennywort, others navelwort. It likes cool, shady places, rock crevices and walls on acid soils. I take it gently out from among the stones, keeping a good bit of soil round the roots, and carry it home to plant in a similar spot in the garden.

PLANTS AND TREES Key to Illustration on pages 60-61

1 Oak 2 Maple 3 Silver birch 4 Lime 5 Elm 6 Wych elm 7 Ash 8 Beech 9 Holly 10 Hazel
11 Hawthorn 12 Buckthorn 13 Bird cherry 14 Yellow archangel 15 Persian speedwell 16 Common forget-me-not 17 Red campion 18 Common violet 19 Shining cranesbill 20 Wall lettuce
21 Butterfly orchid 22 Nipplewort 23 St John's wort 24 Wood spurge 25 Mouse-ear hawkweed
26 Nettle-leaved bellflower 27 Wood avens 28 Enchanter's nightshade 29 Pennywort
30 Bracket fungus

ANIMALS

31 Fallow deer 32 Song thrush 33 Chaffinch 34 Blackcap 35 Nuthatch 36 Hawfinch 37 Blue tit
38 Great tit 39 Wood White 40 Speckled wood 41 Ringlet 42 Comma 43 *Zygiella atrica*
44 Common garden spider 45 *Meta segmentata* 46 Red ant 47 Violet ground beetle 48 Green tiger beetle 49 *Jassus lanio* 50 Forest bug 51 Seven-spot ladybird 52 *Scolytus intricatus* 53 Weevil
54 Grove snail 55 *Cimbex femorata*

The Dawn Chorus

Deciduous woods (those with trees and shrubs which lose their leaves in autumn) have a great variety of habitats for plants and animals. Although different animals live and feed in the same area, they do not compete with each other because each can exploit the wood in a different way. A closer examination of their way of life shows us how many species can live together without squabbles over food or territory. For example, some species are nocturnal, but most are diurnal, i.e. active during daylight hours.

If one enters a wood at sunrise in summer when the dew is still glistening on the spiders' webs, the dawn chorus of birds can be heard. The cock birds are advertising their territories by singing and are warning other males to keep away. If it is early in the breeding season, the singing may also attract a mate. It has been shown by experiments that it is the change in the light intensity which 'triggers' the beginning of the daily activity of birds and many other vertebrates. The daily activity varies slightly between species. For example, by comparing when the first birds sing at dawn it has been found that the nightingale and song thrush are usually earlier 'risers' than the blackcap and goldcrest.

In addition to the intensity of the light, temperature also slowly increases in the early morning and can act as a trigger for the onset of insect activity. Insect muscle functions well only in a narrow range of temperature. If the weather is cool, very few insects will come out of hiding. Birds and other insect feeders in woodland have adapted their daily feeding timetables so that they correspond with the times of the maximum activity of their food supplies.

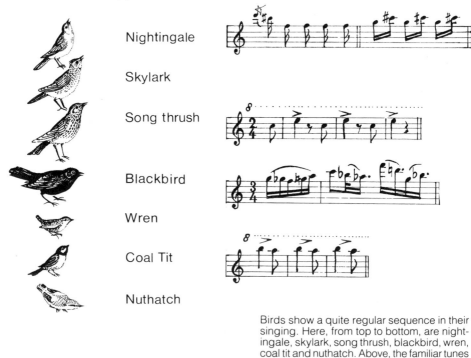

Nightingale

Skylark

Song thrush

Blackbird

Wren

Coal Tit

Nuthatch

Birds show a quite regular sequence in their singing. Here, from top to bottom, are nightingale, skylark, song thrush, blackbird, wren, coal tit and nuthatch. Above, the familiar tunes of some of them.

In a wood, birds feed at different levels. In the highest canopy can be found the wood pigeon and blue tit; lower, from left to right, are the marsh tit, nuthatch, tree creeper and great spotted woodpecker; still lower, the chiffchaff, great tit and blackcap. In the undergrowth are the wren, tree sparrow and nightingale; and at ground level the blackbird.

In a woodland there are many ways of exploiting different resources. Competition among species active by day is avoided by each specializing in a particular way. For example, the nuthatch, tree creeper and great spotted woodpecker all obtain food from the bark of trees. However, the tree creeper has a very slim curved bill, with which it probes into crevices in the bark to get at insects. The bigger woodpecker can dig deeper into the bark and dead wood with its stronger beak, and with its long sticky tongue picks out larger invertebrates which are not available to the tree creeper. The nuthatch is intermediate in size between the tree creeper and woodpecker, and as well as taking some invertebrates from the bark it takes many acorns and other tree fruits, wedging them into a crevice in the bark and hammering them open with its beak. So the three species, although each attacking the bark of trees, are not in direct competition for food and so are able to co-exist in the woodland.

A wood has several layers in it, from ground level to the tops of the trees. The great tit and blue tit, like the tree creeper, nuthatch and woodpecker, are superficially similar in their habits, but avoid competition by using the wood in different ways: the heavier great tit feeds lower down, on the ground or in the shrub layer, and the smaller, lighter blue tit feeds higher in the canopy layers of the trees.

The breeding season of birds is timed so that when the young are in the nest, food is most readily available and the days are long. This ensures that the chicks have the best chance of survival. In the summer, one or both parent birds may have to spend almost all the daylight hours searching for food for their young. One pair of blue tits is capable of collecting several thousand caterpillars each day for their brood. The young are growing very rapidly and need enough food during the day to last them through the night when the parents can no longer collect food and the temperature is lower.

Birds' Nests—how they are built

Birds' nests come in different shapes and are made of various materials, which are shown on the right and also as symbols next to the nests.

Blackbird's nest:

Found in mixed habitats usually under thick cover, sometimes on the ground between the fork of tree roots. The cup is built of grasses and dead leaves with a solid layer of mud and coated inside with fine grasses. The eggs are blue to brown, mottled with light brown.

Magpie's nest:

The magpie breeds in fringes of woodlands in scrub layers. The nest is a bulky cup of sticks bound with mud and plant fibre. It has a loose dome covering with an entrance to the side which is sometimes barbed with bramble or a hawthorn twig to stop predators entering. The eggs are greenish blue, spotted with olive-brown.

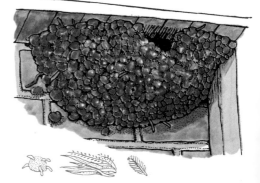

Long-tailed tit's nest:

This nest is found in scrub and hedgerows. Sometimes the tit breeds in trees between the fork of the trunk and boughs. Spiders' webs bind mosses and lichens together forming a camouflaged outside and internally a soft coat of feathers cushions the eggs. The eggs are white, sometimes speckled with light purple spots.

House martin's nest:

Originally the house martin bred in open country under overhanging cliff outcrops. Today nests are mainly found under eaves of houses and buildings. The half cup is attached to a vertical surface (wall) very close to the horizontal (eaves). This leaves only a small entrance protecting the young. The nest is lined with feathers. The eggs are white.

64

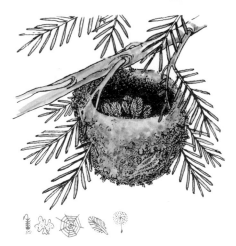

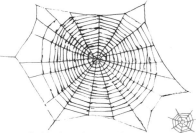

Components of birds' nests

The symbols for these are shown in the margins

Spider's web used for binding

Moss and lichen used for camouflage

Goldcrest's nest:

Found mainly in mixed woods in conifers (yew). The small nest hangs closely under the outer branches which protect it. Spiders' webs bind the cup to its supporting twigs. It is very hard to spot because of its moss and lichen camouflage. Eggs are white, finely speckled with buff-brown.

Feathers used for padding and lining

Thistle down and other plant down

Wool and hair used for padding

Reed warbler's nest:

The reed warbler breeds near to and over water in reedbeds. Sometimes colonies of warblers can be found nesting. The nest is of woven grasses and reeds, lined inside the deep cup with feathers and soft vegetation. The eggs are white speckled with purple blotches.

Other materials used by birds frequently as part of the construction:

sticks, roots, mud, grasses, dead leaves.

The Woodland at Nightfall

There are two reasons why English woods are usually rather lacking in night activity. The first is that, compared with other countries—North America, for instance—we have relatively few medium-sized mammals; while secondly, woods in England are usually interspersed with fields and more open country, which the animals and birds prefer. The Blue Ridge mountains of Virginia, for example, are entirely covered, for hundreds of miles, with a rather dull growth of thin, spindly trees, and here one can reasonably hope to see several kinds of nocturnal animal—woodchucks, skunks, opossums, raccoons, foxes—just possibly a bear. They say that long ago, in the days of slavery, slaves, escaping to the North and walking by night without lights, used to tell their direction by feeling on which side of the tree trunks the mosses and lichens were growing. They grow more thickly on the north, or shady side—or so it was believed. All I can say is that I wouldn't care to have to rely on them for direction.

As many people know, badgers usually dig their earths or setts in deep woodland and come out at night to hunt or play; but badger-watching needs a great deal of perseverance, practice and patience. Badgers are timid, wary creatures and are quite likely to stay underground all night if they have the least suspicion that an intruder is near. A change in the breeze, one clumsy footfall or cracking stick is enough to spoil any hope of seeing a badger, and I have a friend who once stood for three hours without blowing his nose, and let a spider wander all over his neck and ears at the same time.

There is a place in the Isle of Man called Lhergy Dhoo, which in Manx means "the black hillside". Here a little wood, mostly of pine trees, stands on either side of a quiet country road about half a mile from the sea. As night falls, flocks of rooks arrive and circle, tumble and caw in the twilight sky before settling to roost. The wood is also a roosting-place for jackdaws, magpies and the grey-and-black hooded crows which infest the island. Easy to see how the place got its name. No doubt these birds have frequented it for hundreds of years.

To take a sleeping-bag or blankets to the edge of a wood on a warm summer night offers a good chance of hearing or seeing something worthwhile, for someone lying down comfortably and not expecting or waiting for anything in particular is less likely to rustle and to be tense and impatient than a watcher standing up. Many animals seem to be stimulated primarily by smell and sound, and to make relatively little use of their eyes. I remember two hares, playing and tussling with one another, which approached from upwind until I could almost have touched them, when their eyes at last told them what their noses and ears had failed to do, and they dashed away. Smaller creatures, like water-voles or hedgehogs, will carry on their normal activities quite near someone who is perfectly still. Once, in the Blue Ridge mountains, coming home through the woods after sunset (rather like Robert Frost—"Miles to go before I sleep"), I watched two male ruffed grouse (*Bonasa umbellus*) leaping into the air, capering, posturing and apparently trying to intimidate one another on the open ground of a rough track between the trees. What they were up to remains a mystery, for it was autumn and there was no female to be seen. They were quite oblivious of me and in the end I had to interrupt them, for it was nearly dark, I wanted to get home and they were in the way.

Lying awake before dawn on the edge of a wood, one feels the unbroken continuity of time, as the place where one happens to be on the world's surface revolves once more towards the sun and its inhabitants react accordingly. While it is still dark, the owls return from their night's work, calling to one another among the trees. Soon after, the wood-pigeons begin their clattering and cooing and most probably the blackbirds and chaffinches begin the dawn chorus. Moths and slugs disappear and flies and small beetles become active. If there is a stream near-by, the trout are likely to begin rising and perhaps a heron appears to stab and paddle—Dylan Thomas's "elegiac fisher-bird". There's another, different day between about four and eight o'clock on a summer morning—a time of quiet activity before men, with their cars and tractors, raise the noise level.

The nightingale, of course, provides the best reason of all for visiting woods at night. The males arrive in England and Wales during April, usually a week or two before the females. They sing (by day as well as by night) to attract the females, and go on throughout the nesting and breeding season. They are very shy—"habits skulking", as one bird book rather brusquely puts it—and conceal themselves in thick undergrowth. They take a long time to "warm up" and will often go on saying nothing but "Jug, jug" for an hour or more. A nightingale in full song, however, is unforgettable, and more than deserves all that poets have said of him. Here is a sonnet by John Clare, the country labourer who became the most attractive of English nature poets. It is interesting that Clare had evidently had the feeling that he was being teased by the nightingale. So have I. It usually stops singing on your approach and then, just as you're going away, starts elsewhere, further off. (Clare calls the nightingale "she", but nevertheless I think he knew that it is the male who sings.)

> When first we hear the shy-come nightingales,
> They seem to mutter o'er their songs in fear,
> And climb we e'er so soft the spinny rails,
> All stops as if no bird was anywhere.
> The kindled bushes with the young leaves thin
> Let curious eyes to search a long way in,
> Until impatience cannot see or hear
> The hidden music; gets but little way
> Upon the path—when up the songs begin,
> Full loud a moment and then low again.
> But when a day or two confirms her stay
> Boldly she sings and loud for half the day;
> And soon the village brings the woodman's tale
> Of having heard the new-come nightingale.

PLANTS AND TREES   Key to Illustration on pages 68-69

1 Pedunculate oak 2 Beech 3 Silver birch 4 Holly 5 Persian speedwell 6 Nettle-leaved bellflower 7 Yellow archangel 8 Nipplewort 9 Mouse-ear hawkweed 10 Wood spurge 11 Shining cranesbill 12 Wall lettuce 13 Enchanter's nightshade 14 Red campion 15 Butterfly orchid 16 Common violet 17 Wood avens 18 Ivy

ANIMALS

19 Fox 20 Hedgehog 21 Red deer 22 Brown hare 23 Badger 24 Dormouse 25 Yellow-necked mouse 26 White-lipped banded snail 27 Noctule bat 28 Woodcock 29 Barn owl 30 Tawny owl 31 Old lady moth 32 Elephant hawkmoth 33 Yellow underwing 34 Red underwing

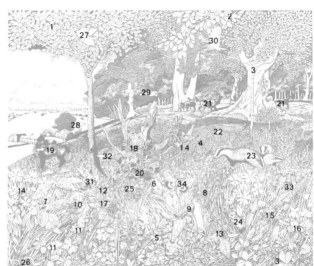

Tracks and Signs

The deciduous woodland, with its oaks and ashes, birches and haw-thorns, its tangly brambles, and its grassy clearings, seems to us so alive and full of birdsong, the rustling of mice, and colour by day. But at night it seems very different. Often the only signs of the animals of the night which we notice are the tracks of foxes, badgers or deer, seen in muddy ground or on sandy areas. We may also find the pellets of tawny owls, which roost in hollow trees by daytime. These pellets are the undigested remains of small birds and mammals which the owl has regurgitated. If you find a pile of them you can be fairly sure that an owl has been roosting in the tree above.

Most woodland plants open their flowers during the day, when there are plenty of insects flying around to pollinate them. But some, such as the evening primrose or the night-flowering catchfly (*Silene noctiflora*)—more usually found in uncultivated fields—rely on night-flying insects for pollination. So their flowers can be seen slowly opening as dusk falls. Plants like these usually have pale flowers and a smell which seems to be particularly attractive to the insects—usually moths—which pollinate them.

Night-flying insects rely very much on scent. In the moist night air, scent hangs for a long time in the atmosphere near the plant, giving the plant more chance of gaining a visiting insect. Insects are in fact very sensitive to smells, and only minute quantities, hardly detectable by man's most complicated scientific instruments, are enough to attract them.

a   b   c   d   e

The presence of animals at night can also be recognised by their droppings.

(a) Rabbit (b) Hare (c) Fox (d) Badger (e) Roe deer

When owls (and hawks) have digested their furry and bony prey, they eject the undigest-ible remains, such as bone, fur, feather in pellets (or casts). By examining the contents of these pellets, it is possible to find out not only the birds' diet, but also the type of small animal or bird living in the vicinity.

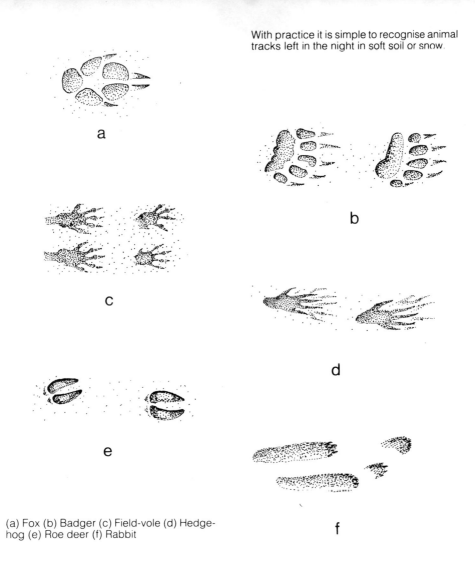

(a) Fox (b) Badger (c) Field-vole (d) Hedge-hog (e) Roe deer (f) Rabbit

Male moths, for instance, can be attracted to a female by her scent over several hundred yards! This sensitivity to scent is vital for ensuring that the two sexes come together for mating.

Most moths fly by night. If a light is positioned in the wood, many thousands of moths will be attracted to it, and these 'light traps' can be used to find out more about moths and their activities over a period of time.

Because they are so plentiful, night-flying moths are important food items for nocturnal predators. During the day, most insect predators are birds, such as fly catchers and redstarts, for instance. By night, the chief enemy of moths is the bat. The commonest bat, the pipistrelle, is well adapted to catching night-flying insects, as is explained in the section on 'Sense and Sound'.

Other well-known nocturnal animals are the tawny owl and the badger. Less known, though familiar to country people, is the woodcock, whose courting display flight, called 'roding', which takes place at dusk and dawn is a thrilling sight. The male bird flies with strange interrupted wing-beats, in a wide circle, uttering two call notes.

Slugs and Snails

Snails and slugs belong to a group of animals called Gastropods—literally 'belly feet'. They are very similar, but slugs lack the snail's spiral shell. Land snails are mainly nocturnal because they would lose moisture during the day.

White-lipped snail *Cepaea hortensis* is common, and there is much variation of colour and marking of the shell within the species.

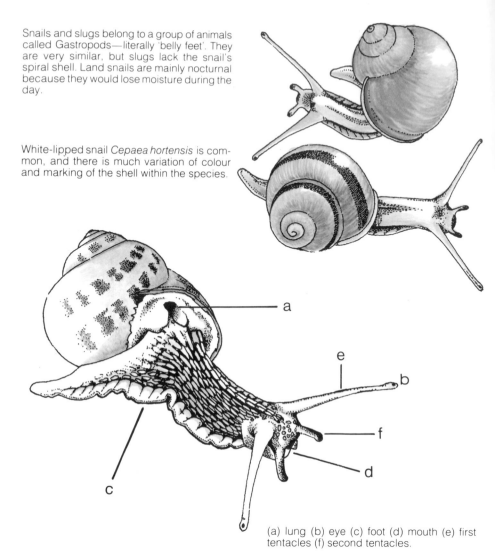

(a) lung (b) eye (c) foot (d) mouth (e) first tentacles (f) second tentacles.

The ramshorn snail is a snail of ponds and streams, very popular with aquarium owners because of its voracious appetite for algae which form on the glass of aquaria.

Amber snails are found in damp meadows and near streams.

In winter many snails hibernate by closing their "door", the operculum, and sealing it with slime.

Snails' eggs are planted in the ground or under stones.

Young snails—their shells are transparent.

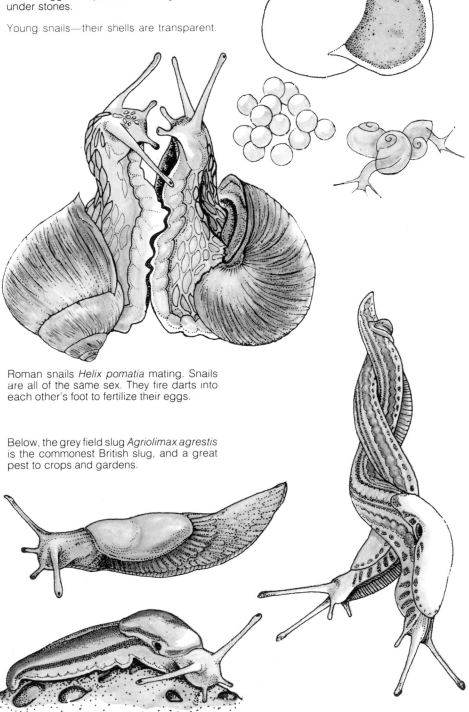

Roman snails *Helix pomatia* mating. Snails are all of the same sex. They fire darts into each other's foot to fertilize their eggs.

Below, the grey field slug *Agriolimax agrestis* is the commonest British slug, and a great pest to crops and gardens.

Above, the wood slug *Arion subfuscus* feeds on fungi, which usually match its colour and so camouflage it.

When mating, the great grey slug *Limax maximus* hangs vertically upside down, entwined with its mate and suspended on a string of slime.

73

A Mountain Stream by Day

I am making a kind of pilgrimage. Starting near the foot of the glen, not far from where the stream runs into the larger river, I mean to follow the banks—and in places probably the actual bed—of the stream, almost up to its boggy, peaty sources on the moor. I'm carrying a short, split-cane rod and I may, as Bertie Wooster would say, drop the casual worm into the odd pool, in case there happens to be a trout on the feed. But I'm not very serious about it. The stream is low. This is a dry August and conditions are all against fishing.

> Bonny Kilmeny gaed up the glen;
> But it wasna to meet Duneira's men . . .
> It was only to hear the yorlin sing
> And pull the blue cress-flower round the spring;
> To pull the hip and the hindberrye
> And the nut that hung from the hazel-tree.

That's about how it is with me, too. The most beautiful thing about a glen stream is its sound, as it chatters and tumbles between and over its slate and granite rocks. The chalk streams of the south country are beautiful too, but for the most part silent. People lucky enough to have grown up in a glen have told me that they miss the sound of the stream when they go away.

Down here, the stream runs through broken woodland that covers the steep hillsides. In places the branches of old, huge trees stretch across from bank to bank—sycamore, beech, Spanish chestnut and ash. Elsewhere, in patches of more open ground, smaller trees flourish among the bracken—grey-green, scrubby osiers (*Salix viminalis*) not more than six or seven feet high, and fuchsia in bloom, long branches of pendent, dark-red flowers hanging over the falls and pools.

Next to the sound of the beck I like the smell of the herbage and woodland along the banks: a moist, leafy smell of dead leaves, wet earth and grass, ferns, lichens and mosses. Surely if there is one family of plants more than another which deserves praise for helping to make this planet what it ought to be, it is ferns. The very word suggests shade, moisture, coolness, verdure, all that a desert has not. There are a lot of ferns here. As well as the common bracken (*Pteridium aquilinum*) in the open spots, there is the male fern (*Dryopteris filix-mas*) with pale-brown scales and clusters of spore cases on the undersides of the tall fronds; the hard fern (*Blechnum spicant*) with its narrow inner fronds and wide-spreading outer fronds of lobes like the teeth of a comb; and—yes, here it is—the Hart's tongue (*Phyllitis scolopendrium*) with its undivided fronds like long, curly, fleshy leaves. The mosses are green and thick even in this August heat—flat carpets of *Eurhynchium praelongum*, all mixed up with the grass and last year's dead leaves; the stiff soldiers-on-parade of *Mnium hornum*, dark-green and rather sombre: the flat mats of *Camptothecium sericeum* growing on the rocks and bases of the trees; and best of all, graceful and delicate as a beautiful miniature fern, *Thuidium tamariscum*.

There's a dipper flying away. I hoped I'd see one—a tubby, dark-brown, white-breasted bird a little bigger than a sparrow. They like hilly districts and fast mountain streams in the west and north; and believe it or not, they can walk under water along the bed of the stream. What happens is that when a dipper walks upstream with its head down looking for food, the force of the current against its back keeps it on the bottom.

As I emerge from the woodland and start climbing the open fell among the heather, the sparse thorn-trees and brilliant-berried rowans, I see a grey wagtail bobbing and strutting on a flat stone under a little waterfall. He is bluish grey above and yellow below, very handsome; but it is his brisk, cocky walk (he never hops) and dipping, flirting flight, and above all the long tail forever wagging like a clockwork toy, that make him look so buoyant and attractive. Beside the stream, in a flat, wet place, rises the slender, foot-long spike of a bog asphodel (*Narthecium ossifragum*)—a thin tuft of orange-yellow flowers on a leafless stem. The long, separate leaves are all hidden in the surrounding grass.

Steeper now, hotter and more arduous. A bush of broom covered with old, crackling, black seed-pods; heather; a curlew calling; and a great hare with black-tipped ears scudding away up the fell. Here's a green, glistening liverwort—*Preissia quadrata*, I think—matting the side of a wet, shady stone; and lichens, golden and grey—*Xanthoria aureola*, and *Cladonia pityrea* with its pretty, scarlet apothecia. Only one kind grows actually in the water, though—*Dermatocarpon aquaticum*—little, separate, roundish, curly-edged flakes, like fairy pancakes.

High up on the fell, not far below the moorland watershed where the stream breaks up into tiny becks and they into oozing, boggy peat-pools and wet grass, there is an old, deserted slate-quarry; and close by, the stream has formed a little tarn (or lough, or pool), no bigger than a lawn tennis-court, but a good twenty feet deep for all that. It's still, silent and shady in the heat, for the western bank is a small precipice, twenty sheer feet of ferny, heathery, gorse-yellow and liverwort-green rock, forever dripping and shining with trickles of water.

I shall stop here, and throw ducks and drakes (or do you call them skimmers?) across the water with the flat slate fragments that cover the shore. It's hardly worth plodding any higher, up to the bogs and the rifted peat. I'll go back by a smoother way, down the shepherd's trod and across by the reservoir. I've followed the stream over two miles in nearly three hours of fairly rough scrambling; and dear me, I quite forgot to do any fishing! Well, if you want to feel like Dr Livingstone, on a hot summer afternoon, this is one way.

> Late, late in the gloaming, Kilmeny came hame.
> Kilmeny had been she knew not where,
> And Kilmeny had seen what she could not declare.
> Kilmeny had been where the cock never crew
> Where the rain never fell and the wind never blew.

It's not at all clear from the poem where Kilmeny *had* been, as a matter of fact. But she'd been up the glen, all right; and it's a beautiful poem—by James Hogg, if you want to get hold of it.

PLANTS AND TREES

Key to Illustration on pages 76-77

1 Ash 2 Sycamore 3 Beech 4 Sweet chestnut 5 Aspen 6 Alder 7 Scots pine 8 Birch 9 Juniper 10 Hawthorn 11 Tea-leaved willow 12 Gorse 13 Blackberry 14 Fuchsia 15 Creeping lady's tresses 16 Bog asphodel 17 Hart's tongue fern 18 Male fern 19 Hard fern 20 *Preissia quadrata* 21 *Thuidium tamariscum* 22 *Camptothecium sericeum* 23 *Mnium hornum*

ANIMALS

24 Bank vole 25 Coal tit 26 Hooded crow 27 Curlew 28 Grey wagtail 29 Siskin 30 Crested tit 31 Goldcrest 32 Linnet 33 Dragonfly 34 Scotch argus butterfly

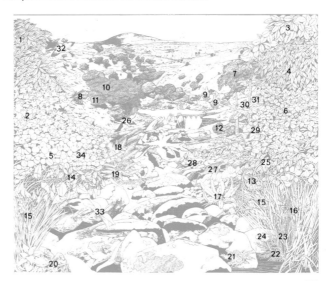

Aquatic Patterns

As has been said, water is more slowly warmed by the sun than is soil. On the other hand it loses its heat more slowly, so there is little difference in temperature in water by day and by night.

Sun also provides light, and it is the sun's light which creates the energy needed by plants to 'photosynthesize'—that is, to combine carbon dioxide in the air or water with their own tissues and give off oxygen. So plants must be active by day, when the sun is shining. But when do aquatic *animals* feed?

The answer is not simple. Many aquatic animals pass through several stages in their lives, and each stage may have different needs. For example, while the beautiful adult dragonflies fly above water, their 'nymph' stages are fierce aquatic predators.

For adults, the most important thing is to find a mate. With dragonflies this depends on sight rather than smell, so mating takes place in daylight. Usually the males arrive in the morning, patrol a sector of the water area, and depart in the evening. The females only visit the area briefly. But at least one species, *Lestes disjunctus*, a damselfly, has maximum numbers at the water by late afternoon, rather than at noon. Its nymphs, however, appear to be more active by night.

Adult amphibians (which include frogs, toads and newts) have a different pattern. They are active at night, and by day hide in holes, or under stones or logs. In the evening they move towards the water and begin calling; what sets them off in the evening is not known. It is probably falling temperature or fading light. But in two species, certainly, the natterjack (*Bufo calamita*), and a tree frog (*Hyla arborea*), their calling is controlled by water temperature. By midnight, when the water has perceptibly cooled, they cease their calling.

If a plant is placed in a corked bottle filled with carbon dioxide and exposed to the sun, the carbon dioxide is slowly replaced by oxygen as photosynthesis takes place.

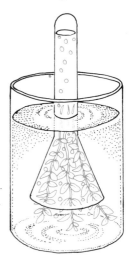

In water saturated with carbon dioxide (e.g. from a soda siphon) pondweed when exposed to the sun will release bubbles of oxygen.

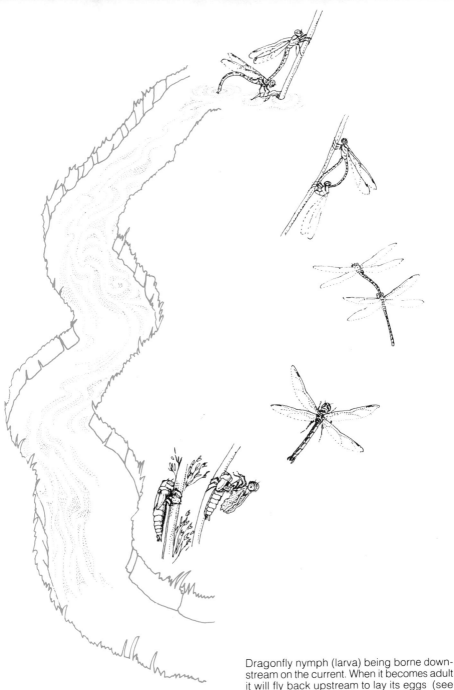

Dragonfly nymph (larva) being borne downstream on the current. When it becomes adult it will fly back upstream to lay its eggs (see page 87).

Fish also show a daily pattern of behaviour. Usually they become more active by night. Sometimes they show different patterns. There are two closely related species of sculpins (broad-headed spiny American fishes), of which one, *Cottus gobio*, is restricted to night-time activity, and the other, *Cottus poecilopsis*, is active at night in summer, but may become active by day in winter, especially if the sky is overcast. This suggests that it is the dimness of the light which makes it active, rather than any internal clock.

Water Insects and their Life Cycles

The larvae of aquatic insects are fierce predators attacking their prey with their powerful
jaws. Some of them are illustrated here.

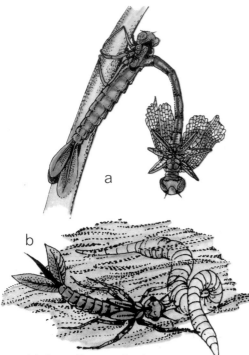

(a) damselfly emerging from pupa.

(b) damselfly larva feeding on worm. This
larva is more slender than that of the dragonfly (see p. 78). It uses its fierce lower lip-mask
to grasp its victim.

Mayfly larvae emerge between March and
May. They are active swimmers and feed on
algae in mud and on stones.

Stonefly larvae crawl about under stones and
on the gravel in streams. They are mainly
vegetarian.

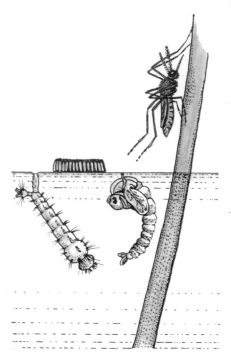

Mosquito. The egg raft floats on the surface,
while the larva (left) and the pupa (right) are
suspended by an air tube just below the
surface.

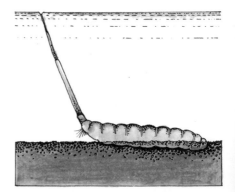

Grubs'. These are legless larvae of flies.
(a) snipe fly larva (*Brachycera*)
(b) crane fly larva (*Nematocera*)
(c) blue bottle larva ('gentle'—*Calliphora*)

The rat-tailed maggot, larva of the drone fly
Eristalis, has a long telescope tube through
which it breathes.

Beetles have four varying types of larvae.

The whirligig beetle larva is a voracious predator. Its mandible has adapted sucking pincers.

The brown china mark moth *Nymphula* has aquatic larvae, which make floating homes from leaf fragments, and safe in these they eat the underside of pond plant leaves.

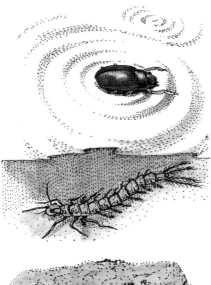

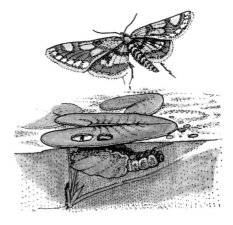

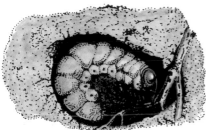

Cockchafer larva takes about three years to develop into an adult beetle. It feeds below the ground on delicate rootlets.

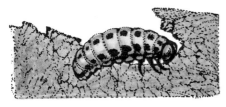

Leaf beetle larva strips plants of their leaves. It has three pairs of powerful legs. The Colorado beetle is included in this group.

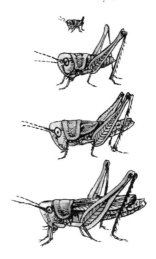

Weevil larvae have no legs, but they have powerful jaws and feed mainly in the roots of trees, causing much damage to forests.

Growth of the field grasshopper *Chorthippus brunneus*. Grasshoppers lay eggs in the ground. The larvae hatch as tiny leggy creatures. They go through several moults and with each moult they become more like adults. The wings are the last part fully to develop.

Still Water at Nightfall

It wants perhaps half an hour to darkness on this stretch of inland water among the Manx hills—a reservoir, half a mile long and two to three hundred yards across. It is getting late in the year—late September, with blackberries on the brambles and mushrooms here and there in the fields on the lonelier, eastern side of the water. Along the Island's western coast, six or seven miles away, it has been a gusty, squally day, with flying cloud, but here, in this narrow valley below the moor, surrounded by hills, it is calmer—calm enough to see the rings made by the fish rising out in the centre of the reservoir. Sea trout run plentifully up the Manx rivers, but it seems just a shade early for them yet; and although I haven't asked the locals, I can't help looking rather doubtfully at the long, steep, smooth slope of the outfall at one end of the dam. Could they negotiate that? Well, anyway, I'm not fishing. High above the water to the north-east rises the long, flat summit of Carraghan—only 1,640 feet, but I've walked up, over and down to get here, thank you very much. The sun is still shining up there, lighting the bracken and heather, while here, over a thousand feet below, night is already almost fallen.

I look over the dam wall at the nearby shore below; and there, among the weeds and scrub covering the exposed upper stones of the reservoir's bed, is a wonderful sight—a large flock of pied wagtails, rather reminiscent of a crowd of excited waiters during the rush-hour. In fact, one of the pied wagtail's other names is "Dishwasher". There must be forty or fifty of them, running, pecking, flirting into the air to catch flies and darting about among the persicary and docks growing along the water's edge. They aren't migrants, but when, as now, they are flocking together in autumn, they are often moving southward. These may well have dropped in from Scotland. Their piping calls, dipping flights and cheerful, happy energy make me think of the bubbles in a great glass of champagne. For a while they alter the whole tone of this rather louring evening.

The heron doesn't, though—he rather intensifies it. He has been fishing—I suppose, though I didn't see him—which he does by wading about in relatively shallow water and stabbing in it with his long beak. It is a shame, but herons tend to get shot on trout streams, which has made our British species as a whole extremely wary and shy, as some of their overseas cousins are not—for example, in Florida. But there is no alternative, really, if you want any trout. A nesting heron with a brood to feed will catch hundreds of trout fry—little trout—in a few days. This heron is knocking off now, as night falls. He has been craftily working under cover, among the scrub willow round the infall of the brook at the upper end of the reservoir, and now he comes flying down the length of the lake with great, heavy wing-beats, head held back on the doubled-up neck and legs stretched straight out behind him. He seems all twilight, dark, slow and a little menacing as he vanishes over the crepuscular, larch-covered slopes to the west.

Besides the wagtails, there are other autumn flocks about, but they are in open fields some way off. I make my way round towards them, for I want to get a look at the redwings. They're usually timid and very quick to be off, but with binoculars I can see them, mingling with curlews, fieldfares and a plover or two as they forage over the green meadows. They look rather like song thrushes, with the same spotted breasts—they are, in fact, members of the thrush family—but the first thing most people notice about them is the conspicuous yellow stripe over the eye, and after that the reddish mark along their sides. They breed a long way north —Iceland and Norway—but come south in flocks about now. They like open fields best. Suddenly, though I'm still a good fifty yards away, I startle this lot and off they all go, showing, as they beat their wings, the red undersides from which they get their name. They fly across the lake and a solitary, brown-dappled redshank, prowling about among the stones, takes off in sympathy, but soon alights again.

It is almost dark as I reach the northern end of the reservoir and start climbing Carraghan again—not over the summit this time, but back across the western shoulder. The lake below gleams silvery-smooth and this though I am walking in a wind that whips across the moor. Strange, isn't it, that when we think of lakes we usually see them, in our mind's eye, at nightfall? Sir Bedivere threw the dying Arthur's sword into the lake at dusk; Keats's knight-at-arms was palely loitering where the sedge was withered from the lake in autumn; and to Yeats, Innisfree's evening was full of the linnet's wings. Lakes, more often than not, appear to us as grave, still, solemn places, surrounded by trees and glimmering at dusk. It's a good example of situation and climate creating an image. Our British lakes lie for the most part between mountains; and here, up in the north of Europe, that means seclusion, shade and often, coniferous trees—Wastwater, perhaps, or Loch Lomond. Further south, towards the Mediterranean, it's different—blue Geneva, or Maggiore dancing in the sun. They seem a long way off tonight.

I come out on the western shoulder and there are two ravens—not rooks or hooded crows or jackdaws, but great ravens—with terrible beaks and feathers like beards at their throats, ripping and quarrelling over some evil meal. They see me and flap heavily away, but before they are lost in the darkness, show up for a moment against the faintly shining, distant water below. Tea will be nice: fire and light.

Key to Illustration on pages 84-85

PLANTS AND TREES

1 Norway spruce 2 Gorse 3 Blackberry 4 Reed 5 Reedmace 6 Purple loosestrife
7 Great water-dock 8 Amphibious persicaria 9 Bog bean

ANIMALS

10 Heron 11 Redshank 12 Raven 13 Curlew 14 Pied wagtail 15 Fieldfare 16 Golden plover

Water Shelters

Most aquatic animals are active by night, thus avoiding predators such as fish or herons. Their night-time activity can be shown by drawing a fine-meshed net through the middle of a stream by day and then repeating the experiment at nightfall. Many more creatures will be found during the dark hours, and with many species there are particularly active periods just after dark and just before dawn.

The seasons can also affect activity. There is a minute two-shelled animal of estuaries called *Hirschmannia viridis*. This is active by night. In the winter it has two peaks of activity at night, but in the summer only one. Is this because the summer nights are shorter, or is it that there is less food about in winter, so that *Hirschmannia* has to make two expeditions to satisfy its hunger? Or is it the internal clock working again? It is hard to tell. Even with animals such as the freshwater shrimp, *Gammarus*, where there are two peaks of activity and two lesser peaks in between, it may be the time the food takes to be digested rather than an internal clock.

There is a caddis, *Potamophylax luctuosus*, which goes through several stages in its life history. First it lives in a case built out of plant material in exposed situations on the bed of the stream; then it transfers its home to a case of stones, and seeks out darker places where the water is flowing more slowly. This caddis is always nocturnal, but in its earlier stages appears to have no internal clock. In later stages the clock is developed, for even in conditions of constant light it continues its cycle of activity and rest.

(a) Mayflies are delicate creatures with fragile transparent wings. Unlike stone flies, they hold their wings vertically over their bodies when resting. When they emerge from the larval skin, they have to undergo a further moult before they are able to fly properly. No other insect does this in the adult stage.

(b) Stone flies inhabit the same sort of habitat as mayflies. They are much more robust than mayflies.

(c) The caddis fly *Phryxanea grandis* (the Great Red Sedge) is one of the largest of its group. It is an insect of lakes and big rivers.

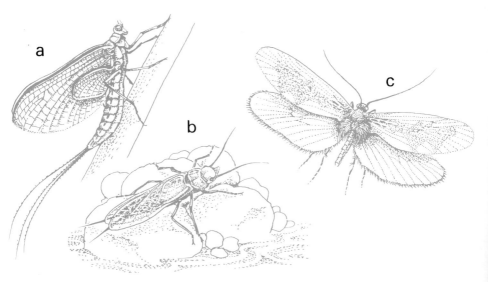

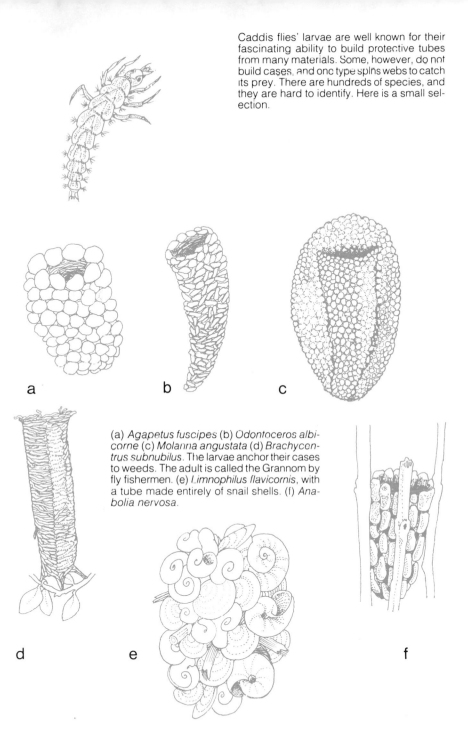

Caddis flies' larvae are well known for their fascinating ability to build protective tubes from many materials. Some, however, do not build cases, and one type spins webs to catch its prey. There are hundreds of species, and they are hard to identify. Here is a small selection.

a

b

c

(a) *Agapetus fuscipes* (b) *Odontoceros albicorne* (c) *Molanna angustata* (d) *Brachycentrus subnubilus*. The larvae anchor their cases to weeds. The adult is called the Grannom by fly fishermen. (e) *Limnophilus flavicornis*, with a tube made entirely of snail shells. (f) *Anabolia nervosa*.

d

e

f

Since so many animals are active at night and get swept downstream by the current, why are any creatures to be found upstream at all? The answer is that the adults can fly, so when they emerge as fully winged 'imagos', they set off upstream to lay their eggs. In this way the upper reaches are constantly replenished with new larvae.

Mosses and Lichens

Mosses, with liverworts, form one of the big divisions of the vegetable kingdom, the Bryophytes. Many are found in damp places, others on walls and rocks.

a stem; b capsule; c spores; d lid; e cap

When the fruit heads are ripe, the 'cap' falls off and the spores are released from the pepperpot-like spore head.

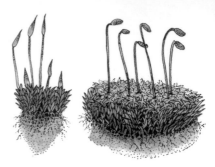

Ceratodon purpureum grows abundantly on high and low land between rocks and walls.

Dicranum scoparium This moss grows in coniferous woods on boulders, stones and dead wood on high land and low.

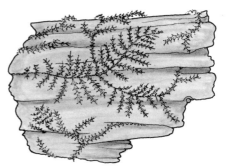

Hylocomium splendens grows in damp forests and on grass slopes over dead wood and vegetation.

Polytrichum commune This grows on mountains, on heathland, along forest streams and in clearings.

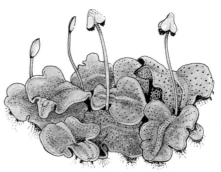

Liverworts are always found in very damp situations. Unlike the leafy mosses, they consist of flat, plate-like "thalli".

Lichens are possibly the most widely distributed of all plants. They are in fact a combination of fungi and algae, living together and dependent on each other.

a) algae cells

b) fungal hyphae

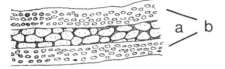

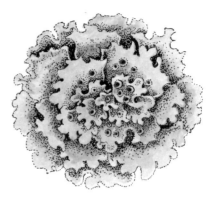

Xanthoria parietina grows on high and low land, on trees, rocks and walls. It forms neat rosettes. Often it can be found near the sea.

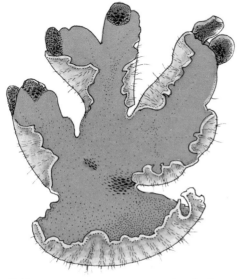

Peltigera grows on moss-covered rocks in open places.

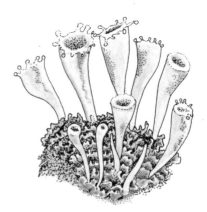

Cladonia fimbriata This grows in woods and heathland on decaying wood and moss-covered rocks usually in quite dry places.

Hypogymnia This forms extensive coatings on tree trunks and branches. Found abundantly on foothills and lowlands.

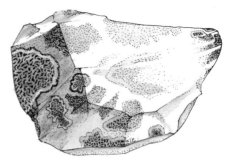

Rhizocarpon geographicum This can usually be found on granite boulders on mountains and exposed places.

A Rocky Coast by Day

There are many kinds of shore, each with its own character and features. The Essex flats, the sandy beaches of Sussex or Dorset, the slate cliffs of Cornwall, the chalk cliffs, the steep, red combes of Devonshire, the wild, rocky coasts of western Scotland—and these are only a few. Beachcombing is always exciting, and even the commoner seaweeds—sea lace (*Chorda filum*), which cannot survive exposure to air for long; bladder wrack (*Fucus vesiculosus*), with its dark-brown fronds and disc-shaped holdfasts; or the striking, crimson *Griffithsia flosculosa*—are well worth admiring. In the old days, fifty years ago and more, children used to collect seaweeds and mount them on cards to make decorative patterns. This is just as enjoyable as ever it was, and so is collecting unusual stones and shells. But may I put in a word here? Don't kill a living creature just to take its shell. When I was in Tahiti, I saw many creatures in the lagoons with large and beautiful shells. Many of the holiday visitors used to dive for them and then get the living creatures out by methods which seemed—and still seem—to me to be cruel, such as boiling them, or burying them in sand for ants to devour. To kill a creature painlessly in order to eat it has always seemed to me to be justifiable. But to take away its life—all it has—just because you want its shell (or skin) for a decoration—well, each of us has to decide for himself or herself what is right and what isn't—but that doesn't seem right to me.

I am sitting on the edge of a high, sheer cliff, overlooking the Irish Sea. It is noon of a very hot day in August. The bees are rambling among the heather and the calm sea lies below like an enormous blue field.

> How fearful
> And dizzy 'tis to cast one's eyes so low!
> The crows and choughs that wing the midway air
> Show scarce so gross as beetles. Half way down
> Hangs one that gathers samphire—dreadful trade!
> Methinks he seems no bigger than his head.
> The fishermen that walk upon the beach
> Appear like mice: and yond tall anchoring bark
> Diminished to her cock; her cock a buoy
> Almost too small for sight. The murmuring surge,
> That on the unnumb'r'd idle pebbles chafes,
> Cannot be heard so high. I'll look no more,
> Lest my brain turn, and the deficient sight
> Topple down headlong.

"Idle" is puzzling at first sight, but here means "useless". Shakespeare certainly captures the curious fear and excitement we all feel in looking down from a high cliff. Also, he evidently knew about choughs—no doubt commoner in his day than ours. On my left, at right angles to where I am sitting, is another cliff, a good three hundred feet sheer to the rocks, and over and round this about thirty choughs are desporting themselves. I am lucky to be watching them, for they are no longer common, being found only along the Welsh, Manx, Scottish and Irish coasts, and then only in certain places. They like cliffs, but seem to feed on the fields at the top rather than down on the shore. Like magpies and rooks, they are members of the crow family, but far more beautiful and attractive in their behaviour. They are black all over, with a glossy, bluish shine against the light, and they have red legs and long, red beaks curving downward at the tip. Their primaries—the big feathers at the ends of the wings—are separated, and this is very noticeable when they fly, for they often sport and play acrobatically in the air in a most delightful way.

The cliff is slate and all the way down its sheer face is broken into deep ledges—

perfect nesting-places for certain kinds of seabird. At this time of year, in August, the ledges, white with droppings and littered with old nests, are occupied by breeding fulmar petrels—beautiful and exciting birds. They have a rounded, chubby appearance about the head and a little black patch, like mascara eye-shadow, round the eye, so that by contrast with the tough-looking, oval-headed, powerfully-billed seagulls they make a soft, rather feminine impression. They swim well but can hardly walk at all, so that as a rule they alight only on the sea or on sheer-faced ledges, from which they can readily take off again. Indeed, they spend most of their life on the wing, gliding smoothly on the air currents with outstretched wings. They breed late—later than all other seabirds—produce only one egg and desert the solitary chick about seven days before it is ready to fly. So it doesn't eat for a week—it has to get rid of its excess fat before it can take off.

From where I am sitting, I can watch the fulmars sailing, sometimes alighting on the ledges and gliding away again, high over the smooth, blue sea; and I can also see the chicks, each sitting alone in its nest on a ledge, very large, wide-eyed and fluffy grey-brown. They'll all be gone in a fortnight—gone to sea, for that is where fulmars live, coming to land only to breed.

As I sit watching, a brown, pointed-winged kestrel comes flying over the fields to the top of the cliff, drops down and alights on a deserted ledge. If it thinks it is going to get peace and quiet it is certainly mistaken, for the choughs begin to "buzz" it, flying close to the ledge, not actually attacking the kestrel but harassing it and letting it see that they don't like it. The kestrel stands it for about twenty minutes and then flies sulkily away again.

I peer over the edge, lying on my stomach, and admire the cliff plants: the pretty sea-campion (*Silene maritima*) with its clusters of dry, white flowers tugging in the wind, many-petalled like rosettes; a few plants of stalked scurvy grass (*Cochlearia danica*), with broad, smooth leaves and small, white flowers, growing in a patch of sand; rock sea spurry (*Spergularia rupicola*), trailing small, pink flowers over the short grass; and, in a cleft, a dry, bushy, close-growing, tough little fern, the sea spleenwort (*Asplenium marinum*)—rather a favourite of mine. What I don't see, half-way down or anywhere else, is any samphire (*Crithmum maritimum*). Never mind, I'll look no more, lest my brain turn. Anyway, the sea, the fulmars, the choughs and the bees in the heather more than make up for samphire.

PLANTS AND TREES   Key to Illustration on pages 92-93

1 Flat wrack 2 Bladder wrack 3 Knotted wrack 4 Sea-lace 5 Channelled wrack 6 Sea couchgrass
7 Sea campion 8 Stalked scurvy grass 9 Sea spleenwort 10 Rock samphire 11 Sea spurry 12 Sea thrift
13 Crested hair grass

ANIMALS

14 Chough 15 Kestrel
16 Peregrine
17 Fulmar petrel
18 Common gull
19 Rock pipit
20 Black-headed gull
21 Beadlet anemone
22 Dahlia anemone
23 Periwinkle 24 Winkles
25 Common whelk
26 Mussels 27 Limpet
28 Common sea urchin
29 Common starfish
30 Spiny starfish
31 Edible crab
32 Hermit crab
33 Prawn (chameleon)
34 Acorn barnacle

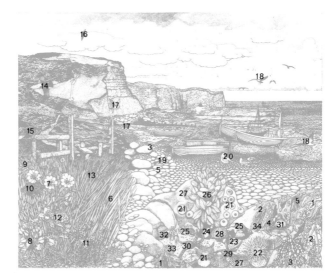

Tides and Internal Clocks

The problems caused on land by daily changes in humidity and temperature do not arise in watery surroundings, such as a pond or a stream—though there is another danger, that of being swept away by the current in running water. But on the seashore, where the tide rises and falls, there is not only the risk of being swept away but also the risk of changes in humidity and temperature. And to make matters worse, tides do not follow a simple day and night rhythm. They are governed by the moon, and their sequence becomes later every day. So we should expect the most complicated 'internal clocks' among seashore organisms, which have not only to tell the time of day but also the state of the tide.

In fact, internal clocks exist even in the tiniest seashore one-celled plants, the diatoms.

Hantzschia virgata is a golden brown diatom that lives in the sand Since it is a plant it must have light to produce its energy. At the same time it must not be carried out to sea by the ebbing tide. This tiny plant can move, and indeed it moves up through the sand to the surface at low tide and then burrows down again as the tide comes in. It has a constant up and down cycle which continues even in a laboratory, under conditions of the same temperature and light. So this diatom must have an internal clock. But it is not a clock which relies on the sun. It is a *moon* clock, and a 'moon day' is about 50 minutes longer than an ordinary day.

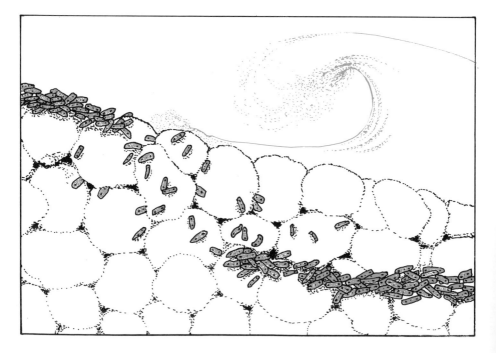

Two-celled plants like the minute diatoms in these illustrations need sun for photosynthesis and lie on the surface of the sand when the tide is out, but when it comes in they burrow about 1/20in into the sand.

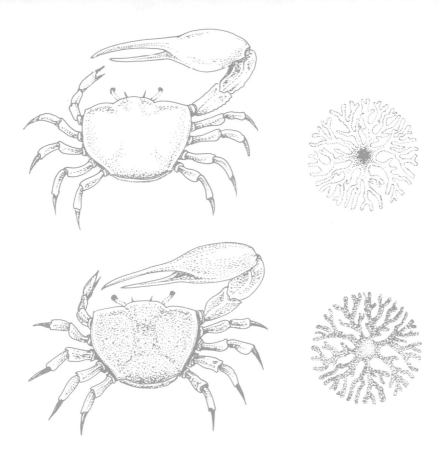

Here the chromataphores in the cells of the crab's shell can be seen contracting into a mass by night and spreading through the cell, thus darkening it, by day.

There is a crab, called *Uca pugnax*, which lives on the beach and shows two peaks of activity—at dawn and dusk. But if it is kept in constant dim light conditions, its pattern of activity gets later by 50 minutes each day, and this is exactly the same as the pattern of the tides. So it seems that it has an internal clock which keeps 'moon time', but which in a natural situation on the beach is 'reset' each day to respond to light and darkness.

Most crabs show a definite pattern of activity. The fiddler crab lives in burrows, and comes out at low tide to feed on decaying matter. As the tide comes in, it returns to its burrow. It would seem that it is the incoming tide which triggers off the clock for this crab, because if it happens to be stranded high up the beach by a pool which only the highest tides, the spring tides, can reach, it reverts to a normal day and night routine, until the next spring tide reaches it.

Crabs also change colour in a regulated way. In their skin are special cells with grains of colouring matter in them. When the grains are tightly packed, in the centre of the cell, the skin appears light; but when the grains are spread through the cell, the skin becomes darker. These crabs are usually light at night and dark during the day; but the fiddler crab also has the pattern of the tides imposed on the day and night skin colour changes; at low tide it grows even darker.

Plant and Animal Life of the Seashore

Jetsam is anything washed up on the beach. Often seen are cuttlefish bones (used for feeding caged birds), the mulberry-like shells of the whelk's eggs, and the skate's egg case, or 'mermaid's purse'

The red starfish feeds on mussels and other shell fish, which it can prise open with its powerful suckers.

The acorn barnacle is extremely common on rocks. When the tide is in, its tentacles emerge from the shell to gather food.

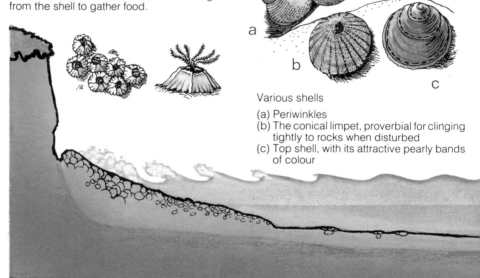

Various shells

(a) Periwinkles
(b) The conical limpet, proverbial for clinging tightly to rocks when disturbed
(c) Top shell, with its attractive pearly bands of colour

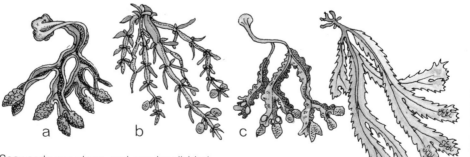

Seaweeds are algae, and can be divided into brown, green and red groups.

(a) Channelled wrack *Pelvetia canaliculata*
(b) Knotted wrack *Ascophyllum nodosum*
(c) Bladder wrack *Fucus vesiculosus* (d) Serrated wrack *Fucus serratus*

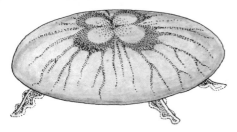

The moon jellyfish *Aurelia aurata* is easily stranded by the outgoing tide, and then soon melts and shrivels to nothing.

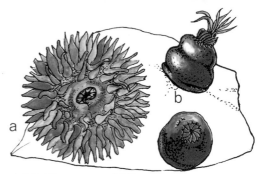

(a) Dahlia anemone (b) Beadlet anemone

The beautiful sea anemones attract passing fish with their waving tentacles. These are tipped with sting cells, which semi-paralyse their prey.

The hermit crab wears marine gastropod shells (in this case a whelk) to protect its soft abdomen.

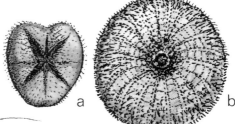

(a) Heart urchin (b) Common urchin

The common shrimp *Crangon vulgaris* is one of the commonest crustaceans, often visiting shallow water in vast shoals, though it is also found in deep water.

Sea urchins are related to starfish. If you imagine a starfish with all its legs sewn together, you have something rather like an urchin. Urchins are covered with spines, and between the spines are a range of differently equipped tentacles, which convey food to the mouth, on the underside of the urchin.

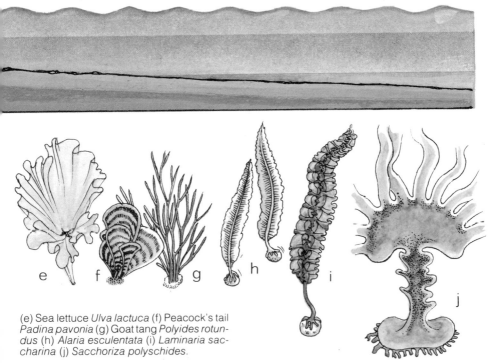

(e) Sea lettuce *Ulva lactuca* (f) Peacock's tail *Padina pavonia* (g) Goat tang *Polyides rotundus* (h) *Alaria esculentata* (i) *Laminaria saccharina* (j) *Sacchoriza polyschides*.

97

The Shore at Dusk

I am by the sea again, but this time it is dusk, and a very different shore from the steep cliffs where the fulmars nest and the choughs play. This is as lonely a beach as one could well find throughout the British Isles—now, at low tide, a hundred yards of flat, wet sand, extending from the shallow froth of the water-line up to the marram grass on the hillocky dunes. It glistens empty for mile after mile—nothing but the coastguard's lookout hut (flying the blue ensign) and a passing fishing-boat a long way out. It is early September, and out at sea the huge, red sun has almost touched the water. It reminds me a little of that unforgettable scene towards the end of H.G.Wells's *The Time Machine*, when the hero, having ridden the time machine to the end of the world, finds himself alone among giant lobsters and more mysterious, half-seen creatures, on a perpetually twilit shore, on which the dark-red, dying sun never rises more than a few degrees above the horizon.

I look round for a little reassurance, and there it is, perched on a strand of barbed wire—a stonechat, most cocky and cheerful of little birds, about as big as a robin, very trim with his jet-black head, chestnut breast and white patches. He bobs and flirts on the wire and makes little, fluttering flights above it and back. He's not terribly common, but usually found near the sea, and he tends to move south in autumn.This one's fly-catching, and very busy getting a full crop before darkness falls. I move on over the dune and begin walking quietly along the landward edge of the beach. The sound of the waves will cover my tread as I approach the outfall.

The outfall is nothing much, actually—just a brook that runs out of the heathery, marshy meadows inland, and down over the sand and pebbles. Like nearly all outfalls, however, it attracts seabirds looking for food, and there they are, sure enough, just settling down for the night, all standing or sitting with their heads to the wind. (They don't like their feathers ruffled.) Most conspicuous are the oyster-catchers—handsome, black-and-white birds about as big as ducks, with pink legs and rather long, orange beaks. Some of them are still pecking about among the stones—they eat all sorts of things, not only oysters. They're rather timid, and as I approach they take fright and fly off along the shore, their black-and-white wings beating rapidly as they utter their high, piping alarm notes. The gulls, however, stay put. Most of them are common gulls, but there are some black-headed gulls, too—their black heads moulting now for winter—and some great black-backs, which I always admire, if only because they're such big, ugly brutes, with fearsome beaks. They tend not to flock like the common gulls, but you quite often see two or three of them feeding or roosting among other shore birds.

Out at sea, more exciting seabirds are fishing in the last light—again, trying to get a full crop to keep them warm through the hours of darkness. Four or five terns are flying up and down, hovering over the water with their beautiful, slender wings and forked tails outstretched, and dropping down to peck and flutter in the waves. (They very seldom submerge.) Arctic terns are the most astonishing migrants of all. Every year they fly south from Canada, Greenland and northern Europe, down the west coast of Africa to the Cape of Good Hope and so on to the Antarctic continent. They make a round trip of 24,000 miles during the eight months of the non-breeding season, averaging 100 miles a day. So I salute these particular terns with awe and a little envy too. They'll soon be gone—they're only in passage here.

In the failing light, I can just see a gannet flying southward, fast, low over the water and some way out. He is easy to identify because of his rapid wing-beats and black-tipped wings. Suddenly he folds back his wings and plunges like a bullet, disappearing for several seconds. As he comes up, another fisherman comes into view in my binoculars—this time a cormorant. (Or it could be a shag—the light's so bad now.) He is all black, flying even lower over the water, with rapid wing-beats and neck outstretched like a swan. I am always rather puzzled by these straight-line, purposeful flights of cormorants. Where do they think they're going? They're not migrants, and as for fishing, they do that like ducks, by swimming and diving. When they fly like this they're not fishing—you never see them dive from flight like a gannet. So they're shifting ground, going somewhere, and far more purposefully and directly than most other seabirds. Anyway, on he goes and out of sight, very intent on wherever it is—perhaps a rock to roost. They prefer rocks.

As I lower my glasses, I realise that quite close to me—not thirty yards away—there are some ringed plovers among the stones. These are most attractive little birds, no bigger than baby chickens, with black-and-white faces and black "bibs". You generally come across them in small flocks of about four to twelve, pecking about on sandy, pebbly beaches, especially towards sunset. Among stones they are so perfectly camouflaged that one can easily fail to see them, even at short range; but as if to make up for this, they will generally allow you to get quite close to them—I suppose because their instinct is to rely on their protective colouring rather than on flight. They look like toys and seldom seem afraid.

The sun is below the sea now, and in the heavy dusk three curlews go flying along the waterline, wheel and turn inland over the sand-dunes and scrubby meadows. Curlews often feed along with gulls on the shore, but during high tide they return inland. Every time I see a curlew I'm struck again by the astonishing length of the great, downward-curved beak—about half as long as the bird itself from head to tail. However, as I'm watching these I suddenly catch sight of something still more exciting. It's a seal, about fifty yards out in the breaking waves. He surfaces, head and shoulders, looks all round, then submerges for some time and reappears a little way further south. This isn't seal coast—grey seals like rocks as a rule. He must be in passage down this lonely, shallow, sandy shore. Seals have been ringed in the Shetlands and recovered off Portugal before now.

I make my way back in the falling darkness, feeling very land-bound and stay-at-home. *I'm* not a creature travelling hundreds—thousands—of miles by flying or swimming. All the same, I've just had some fine glimpses of several who are; and I wish them luck. Nowadays they need it more than ever. I won't tell you about the razorbill I found blinded by oil.

Key to Illustration on pages 100-101

PLANTS AND TREES

1 Sea couchgrass 2 Lyme grass 3 Sea rocket 4 Sea holly 5 Marram grass 6 Sea spurge 7 Sea kale 8 Serrated wrack 9 Caragheen 10 *Laminaria digitata*

ANIMALS

11 Porpoise 12 Cormorant
13 Gannet 14 Manx
shearwater 15 Eider
16 Common scoter
17 Red-breasted merganser
18 Ringed plover
19 Oyster-catcher 20 Great
black-backed gull
21 Common gull
22 Black-headed gull
23 Common tern
24 Razorbill 25 Guillemot
26 Mermaid's purse
27 Moon jellyfish
28 Common cockle
29 Curved razor shell
30 Common urchin
31 Heart urchin
32 Common shrimp
33 Hermit crab

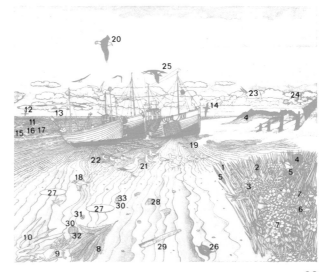

The Moon, the Shallows and the Deeps

Most of the smaller seashore creatures are nocturnal. In daytime they run the risk of being dried out by the sun, or pounced on by predators. The sea slater (*Ligea oceanica*), for example, which is up to an inch or so long, is a tasty morsel for a crab or a gull. At night, when the tide is out, it emerges from its shelter above the high-tide mark to feed on seaweeds, such as *Fucus* (the bladderwracks) and *Pelvetia*. It is extremely shy of light; even bright moonlight will stop it moving. So perhaps its activity is switched on only when darkness falls. But we do not know how it can tell when the tide is out.

A related animal, the sand hopper (*Excirolana*), ventures out only at *high* tide. It emerges from the sand to swim and feed in the breaking wavelets, but as the tide ebbs, it burrows into the sand again. This pattern of activity will continue even without the ebb and flow of the tide. So the sand hopper must have an internal clock.

These daily patterns of activity are also found in the shallows, where the shrimp *Crangon vulgaris* burrows by day and comes out at night. The masked crab (*Corystes cassiovelaunus*) does the same thing.

Presumably these crustaceans have taken to the night to avoid hungry birds. But if the birds hunt by sight during the day, what happens in the short winter days when tides happen to cover the shore? The answer is

a

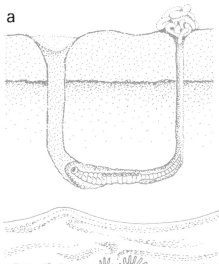

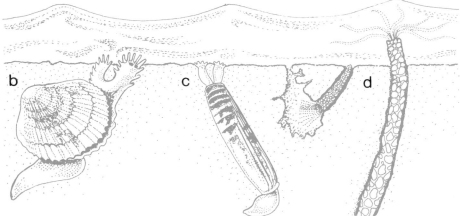

a) The lugworm can easily be detected by the 'casts' it makes on the beach and by the hole at the other end of its U-shaped tube.

b) The common cockle buries itself in the sand to a few inches when the tide is out.

c) The razor shell lies vertically in sand. If disturbed, it can draw itself down to as much as a foot below the surface.

d) Bristle worms are another common seashore creature, also burrowing in the sand.

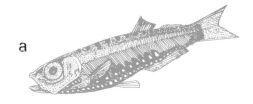

a

Many marine animals are luminescent, especially those in the deepest sea.

The hatchet fish *Argyropelaeus* (a) emits light from the underside of its body.
The pearl side *Maurolichus* sp. (b) and the viper fish *Chauliodus* (c), an angling fish, also emit light from spots on their sides.

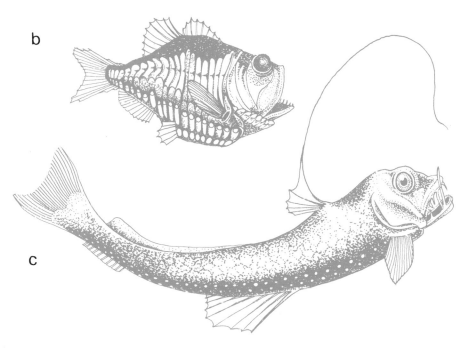

b

c

that they can also feed by night, but often change their hunting techniques. The redshank, for example, pecks its food from the sand or mud during the day, but at night it moves along swinging its bill to and fro in the mud until it contacts its prey. With birds like this, the need for food has overruled the internal clock. A day bird has temporarily been changed into a night bird.

Not all organisms are active only for limited periods. There is a tiny one-celled animal called *Gonyaulax* which, like a plant, needs sun to photosynthesize. But at night it glows, and moves in response to stimuli and also during reproduction.

It is not known why *Gonyaulax* glows. Glowing (or 'luminescence') generally seems to be a form of communication, developed where other systems do not work. This happens particularly among creatures of the deep seas. With some deep-sea fish, luminous organs act as a bait to lure their prey; some squids squirt out luminous fluid to confuse their pursuers. On land, luminescence is generally a way by which some insects, notably fireflies and glow-worms, signal their presence to a mate.

INDEX

Boldface page numbers refer to the Keys for the large landscape illustrations. *Italic* page numbers refer to other illustrations and to captions. Roman numbers refer to text.

hairy stonecrop, *Sedum villosum,* **51**
hairgrass, *Deschampsia spp.,* 46
hard fern, *Blechnum spicant,* **19, 51,** 74, **75**
hare, *Lepus capensis,* 66, 70
hart's tongue fern, Phyllitis scolopendrium, 74, **75**
hatchet fish, *Argyropelaeus,* 103
hawfinch, *Coccothraustes coccothraustes,* **59**
hawthorn, *Crataegus monogyna,* **59,** 70, **75**
hazel, *Corylus avellana,* **59**
heart urchin, *Echinocardium cordatum,* 97, **99**
heath bedstraw, *Galium saxatile,* **43**
heath rush, *Juncus squarrosus,* **51**
heather, *Calluna vulgaris,* **19,** 46, 74, 82
hedgehog, *Erinaceus europaeus,* 26, **27,** 30, 66,
   66, *71*
Hercules, **35,** *40*
hermit crab, *Pagurus bernhardus,* **91,** 97, **99**
heron, *Ardea cinerea,* 82, **83,** 86
Hirschmannia viridis, 86
holly, *Ilex aquifolium,* **59, 67**
honey fungus, *Armillaria melea,* 32
honeysuckle, *Lonicera periclymenum,* **11**
hooded crow, *Corvus cornix,* **75**
house martin, *Delichon urbica,* 64
hoverfly, *Volucella zonaria,* 49
hunting spider, *Pisaura,* 49
Hydra, *40, 41*

internal clock, 15, 22, 30, 38, 47, 79, 86, 93, 102
Irish heath, *Erica hibernica,* 42
ivy, *Hedera helix,* **11, 27, 67**

jackdaw, *Corvus monedula,* 42, 66
Jassus lanio, **59**
juniper, *Juniperus communis,* **75**

kestrel, *Falco tinnunculus,* 15, **43,** 54, 56, **91**
knotted wrack, *Ascophyllum nodosum,* **91,** 96

lady's smock, *Cardamine pratensis,* **11**
Laminaria digitata, **91**
larch, *Larix,* 58
leaf beetle, *Chrysomelidae, 81*
Leo, 34, *40, 41*
Lepus, *41*
Libra, *40, 41*
lichen, **51,** 75, **89**
   Cladonia fimbriata, 89
   Cladonia pityrea, 75
   Dermatocarpon aquaticum, 75
   Hypogymnia, 89
   Peltigera, 89
   Rhizocarpon geographicum, 89
   Xanthoria aureola, 75
   Xanthoria parietina, 89
lime, *Tilia vulgaris,* **59**
limpet, *Patella vulgata,* **91,** 96
ling, *Calluna vulgaris,* 42, **43,** 47
linnet, *Acanthis cannabina,* 10, **11,** 42, **75,** 83
little owl, *Athene noctua,* **11, 27**
liverwort, 75, 88
   Preissia quadrata, **75**
lizard, *Lacerta spp.,* 16
long-eared bat, *Plectos auritis,* 26
long-tailed tit, *Aegittalos caudatus,* 64
lousewort, *Pedicularis sylvatica,* 42, **43**
lugworm, *Arenicola marina,* 102
luminescence, *32-33, 103*
Lupus, *41*
lyme grass, *Elymus arenarius,* **99**
Lyra, 34, **35,** *40*

magpie, *Pica pica,* 42, *64,* 66, 90
male fern, *Dryopteris filix-rnas,* 74, **75**
mallard, *Anas platyrhynchos,* 39
Manx shearwater, *Puffinus puffinus,* **99**
maple, *Acer campestris,* **59**
marram grass, *Ammophila arenaria,* **99**

Mars, **35**
marsh andromeda, *Andromeda polifolia,* **43**
marsh gentian, *Gentiana pneumonanthe,* **51**
marsh marigold, *Caltha palustris,* **11, 27**
marsh tit, *Parus palustris,* 63
masked crab, *Corystes cassiovelauscus,* 102
mat grass, *Nardus stricta,* 46
mayfly, *Ephemera vulgata,* **11,** *80, 86, 89*
meadow brome, *Bromus commutatus,* 30
meadow fescue, *Festuca pratensis,* 30
meadow grass, *Poa pratensis,* 30
meadow sweet, *Filipendula ulmaria,* **11**
merlin, *Falco columbarius,* 54
mermaid's purse, **99**
Meta segmentata, **11, 59**
midge, fam. *Ceratopogonidae,* 15, 30
migration, **39,** 42
Milky Way, *41*
millipede, fam. *Diplopoda, 30-31*
Mira, *40, 41*
Misumena vatia, **11**
moon, 9, 34, 94
moon jellyfish, *Aurelia aurata,* 97, **99**
moorhen, *Gallinula chloropus,* **11**
mosquito, ord. *Culicidae, 80*
moss campion, *Silene acaulis,* **51**
mosses, *Bryophytes*
   Camptothecium sericeum, 74, **75**
   Ceratodon purpureum, 88
   Dicranum scoparium, 88
   Eurhynchium praelongum, 74
   Hylocomium splendens, 88
   Mnium horum, 74, **75**
   Polytrichum commune, 88
   Thudium tamariscum, 74, **75**
mossy saxifrage, *Saxifraga hypnoides,* **51**
moths, *54, 67, 70-71*
mountain hare, *Lepus timidus,* **43, 51**
mouse-eared hawkweed, *Pilosella officinarum,*
   59, 67
mussel, *Mytilus edulis,* **91**
mute swan, *Cygnus olor,* **11**

natterjack, *Bufo calamita,* 78
nests, *64-65*
nettle-leaved bellflower, *Campanula trachelium,*
   59, 67
night-flowering catchfly, *Silene noctiflora, 17,* 70
nightingale, *Luscina megarhyncha, 62, 63,* 66
nimbostratus, **19**
nipplewort, *Lapsana communis,* **59, 67**
noctule bat, *Nyctalus noctula,* **67**
Norway spruce, *Picea abies,* 58, **83**
nuthatch, *Sitta europaea,* 58, *62, 63*

oak, *Quercus robur,* **27,** 58, **59, 67,** 70
occlusion, *25*
old lady moth, *Mormo maura,* **67**
operculum, *73*
Ophiuchus, *40*
orange hawkweed, *Pilosella aurantiaca,* 10, **11**
Orion, 34, *40, 41*
otter, *Lutra lutra,* **11**
owls, 55, *56, 67,* 70
ox-eye daisy, *Chrysanthemum leucanthemum,*
   10, 11, 17
oyster-catcher, *Haematopus ostralegus,* 98, **99**

peacock's tail, *Padina pavonia,* 97
pearl side, *Maurolichus sp.,* 103
Pegasus, *40, 41*
pennywort, *Umbilicus rupestris,* **59**
peregrine, *Falco peregrinus,* **91**
periwinkle, *Littorina littoralis,* 91, 96
Perseus, 34, *40*
Persian speedwell, *Veronica persica,* **59, 67**
Phoenix, *41*
photosynthesis, *78, 94, 103*
pied wagtail, *Motacilla alba,* 82, **83**